Auto Upholstery & Interiors

A DO-IT-YOURSELF, BASIC GUIDE TO REPAIRING, REPLACING OR CUSTOMIZING AUTOMOTIVE INTERIORS

BRUCE CALDWELL

HPBooks

HPBooks
A division of Penguin Group (USA) Inc.
375 Hudson Street
New York, New York 10014

First edition: April 1997

© 1997 Bruce Caldwell
22 21

Library of Congress Cataloging-in-Publication Data

Caldwell, Bruce.
    Auto upholstery & interiors: a do-it-yourself, basic guide to repairing,
replacing or customizing automotive interiors/Bruce Caldwell.—1st ed.
       p.   cm.
    Includes index.
    ISBN 978-1-55788-265-3
    1. Automobiles—Upholstery—Amateurs' manuals.   I. Title.

TL256.C35   1997   96-44793
629.27'7—dc20   CIP

Book design & Production by Bird Studios
Interior photos by the author unless otherwise noted
Cover photos by the author

All rights reserved. No part of this publication may be reproduced, stored in a retrieval system, or transmitted in any form, by any means electronic, mechanical, photocopying, recording or otherwise, without the prior written permission of the publisher.

NOTICE: The information in this book is true and complete to the best of our knowledge. All recommendations on parts and procedures are made without any guarantees on the part of the author or The Berkley Publishing Group. Tampering with, altering, modifying or removing any emissions-control device is a violation of federal law. Author and publisher disclaim all liability incurred in connection with the use of this information.

ACKNOWLEDGMENTS

I would like to thank the many people who helped make this book possible. Models were needed for the photos and in cases where I was the model, someone had to man the camera. The prime models and photographers' assistants were my wife, Denise, and our children, Brent, Bryan, and Brooke. Steve and Sean Wallner were very patient about the extra time it takes to restore a Mustang interior when there are so many interruptions for photographs. Gary Cler took time from his busy upholstery shop, Trimcraft Custom Upholstery in Snohomish, Washington, to appear in photos and answer countless upholstery questions. Brian Kennedy also appeared in many photographs and shared his professional expertise as the manager of the Wild West Mustang Ranch in Woodinville, Washington. Michael Lutfy demonstrated incredible patience and perseverance in editing this book and trying to keep me on some semblance of a schedule. Van Nordquist of Photo Graphic Designs in Mukilteo, Washington, did all the fine film developing and made the prints used in the book.

Several fine companies assisted with products used in the photos. C.A.R.S., Inc., of Berkley, Michigan, and Fullerton, California, was my source for excellent classic Chevrolet reproduction parts and supplies. C.A.R.S. also opened their manufacturing facility for photos of how upholstery kits are made. Stinger Electronics in Clearwater, Florida, supplied insulation material. Restomotive Laboratories POR-15, Inc., in Morristown, New Jersey, supplied rust preventative products. The Eastwood Company of Malvern, Pennsylvania, was my source of affordable upholstery tools and supplies.

Thank you to everyone who helped make this book a reality. ∎

Contents

	Introduction	v
1	Planning Your Project	1
2	Tools & Supplies	13
3	Painting Interior Parts	26
4	Replacing Headliners	34
5	Carpet	43
6	Door Panels	52
7	Dashboards	67
8	Seat Upholstery Kits	76
9	Kick Panels & Package Trays	93
10	Trunks	99
11	Steering Wheels	105
12	Custom Seats	112
13	Dyes & Color Changes	117
14	Minor Repairs & Detailing	122
	Index	134

INTRODUCTION

Automotive upholstery is easy. There, the secret is out. The majority of automotive enthusiasts tackle their own mechanical repairs. Rebuilding an engine is a complex task, but most moderately experienced home mechanics will readily take on a rebuild. Many raw beginners succeed at engine building, but the mention of auto upholstery intimidates them. That doesn't need to be the case.

Granted, complete start to finish custom upholstery jobs are best left to experienced craftsmen, but a great many upholstery projects can be handled by even the most inexperienced hobbyists. The fact that custom upholstery is an art based on lots of technical expertise seems to transcend all forms of upholstery, but that is only the myth, not the truth. The truly tricky work usually involves just the most highly stylized custom work or the restoration of rare classic automobiles. Below these high profile examples are quite a few simple repairs, upgrades, and recover jobs within the reach of a novice enthusiast with limited skills and time. There is absolutely no reason to think that you can't successfully handle most basic upholstery repairs and restorations.

The scariest part of upholstery is sewing. If you're like me, you probably balk at sewing a button on your shirt. But the great news about this book is that it doesn't matter if you can sew, weld, or do complex logarithms in your head. There is absolutely no sewing required for the upholstery tasks covered in this book. The automotive hobby has become so sophisticated that there are pre-sewn upholstery kits for most popular cars and even an increasing number of trucks. That means that all the difficult work is done for you. All you have to do is remove the old upholstery, make any necessary repairs to the underlying structures, and install the new pre-sewn upholstery.

The goal of this book is to teach you a lot of little things that will add up to the ability to restore the interior of your car or truck. Hopefully, you can find information of value wherever you look in this book. Since the information presented here doesn't deal with sewing, the procedures aren't necessarily linear. That is, there are certain tasks that make sense to do before other jobs, but there aren't many hard and fast rules.

You can restore the items that need it most first, or fix the things that are most affordable. To completely gut an interior and restore all aspects of it is a great way to go if you have the time and money, but it is also perfectly acceptable to restore an interior a little bit at a time.

It is very difficult to rebuild an engine one piston at a time, but you could opt to recover the front seats now and wait a couple months before

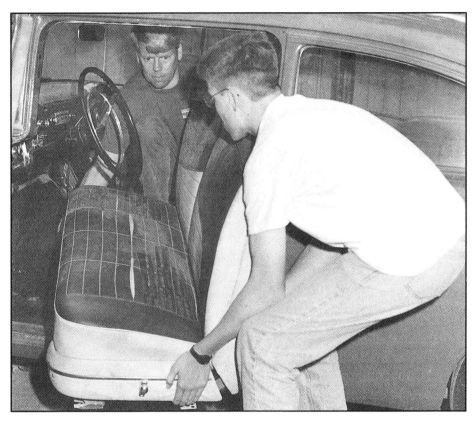

Automotive upholstery is a lot easier than most people think. Much of the work involves removing the old components and replacing them with new, reproduction parts. Most of the tasks can be performed by one person, but it is always nice to have some help. My wife and kids helped in the restoration of our '56 Chevy.

doing the back seat. The fact that so many interior restoration tasks are easy to separate from other operations makes this type of easy auto upholstery perfect for weekend projects. With the projects detailed in this book you can avoid having your car out of commission for any great length of time.

The photos in this book were taken on a variety of cars, but the techniques are similar for most vehicles. Different vehicles may have a few unique fasteners or assembly sequences, but the basic techniques are very similar. It is a good idea to obtain a factory shop manual for your particular vehicle. These detailed books are good for all types of repairs and maintenance. The less expensive, smaller repair manuals that are sold in auto parts stores and book stores don't go into as much detail as the factory shop manuals. These repair manuals are usually condensed versions of the big factory books. You can't always bank on the specific information you need being in these books, but they are much better than no manual at all. If you have a very popular car like a classic Chevy, a Mustang, a Camaro, or a Corvette, chances are good that there is a specialized restoration book (or books) available for your car. These books are great for learning about specific details of your vehicle.

Besides using photos of different cars, the photos aren't always in sequence. Some things were done slightly out of sequence to facilitate the photos and other times the people working on the cars just did the work in a random manner. Upholstery isn't like torquing cylinder heads—it is very difficult to seriously foul up the sequence of events when installing upholstery kits. For example, it generally makes sense to install a new headliner while the seats are out of the car, but you can also replace the headliner with the whole interior intact. Even tasks like painting interior parts can be done with the new upholstery in place; you just have to spend more time masking the area to protect the interior from overspray.

Just as the sequence of events isn't critical with most upholstery tasks, there is often more than one way to do a job. Different people prefer different techniques and tools. This can be confusing to first timers, but our advice is to learn as much as you can and then experiment. A classic upholstery variance is the use of straight or bent nose hog ring pliers. There are pros who swear by each design. After installing some hog rings, you will probably discover which tool best suits your physical needs. Both tools will get the job done, but one style may be faster or more comfortable for a person who does this work all day long. The point is not to worry about the nuances; the little details will take care of themselves as you become more experienced and confident.

When you can improve your car, enjoy the improvement every time you drive the car, and add to the value of the car when you sell it, that is an excellent investment. That is the great thing about automotive upholstery. A high performance engine, a flashy paint job, or expensive custom wheels will easily impress friends and strangers, but a nice interior will make your life nicer every time you drive the car. Making a choice for substance over flash isn't an easy decision for most car enthusiasts, but it is a decision you will appreciate once you make it. Of course, the ideal situation would be a car with a great interior cloaked with a beautiful body and powered by a super powerplant.

I have enjoyed learning about automotive upholstery and I get a lot of satisfaction every time I repair, restore, or improve the interior of one of my cars. I hope this book will afford you the same pleasures. ∎

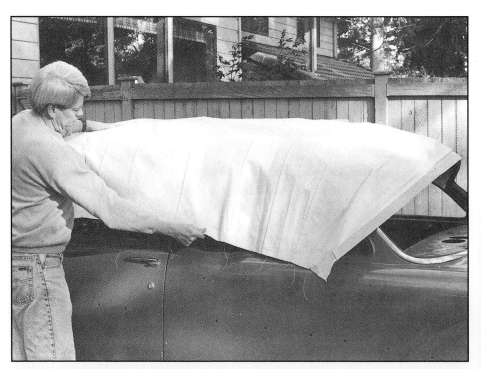

The fact that all the hard work is done by the reproduction upholstery companies makes even somewhat daunting jobs like headliner replacement within the grasp of beginners. The headliner kits are exact duplicates of the factory headliners. All the sewing is done at the factory.

Planning Your Project

1

Plan now or pay later. It is an old adage, but one worth heeding. Upholstery mistakes are not likely to be as severe as bodywork or engine building mistakes, but they can still waste a lot of time and money. By planning your upholstery projects thoroughly, the job will go much smoother, with fewer mistakes.

The main pitfalls of poor planning are re-doing previous work, damaging recently restored upholstery, and making the job more difficult than it needs to be. Planning also helps you focus on the true goals for the vehicle. It doesn't make financial sense to put a show quality interior in a car you plan to sell in the near future. You also need to decide whether it makes more sense to restore an interior to stock, to modify the interior (mildly or radically), or to find a blend of styles that best suits you.

Mistakes Will Happen

If you do make some mistakes, don't feel badly because you are not alone.

I once had an incredibly beautiful luxury seat built for a modified '72

When you have a thrashed interior like this Camaro, it is pretty obvious that the seats and carpeting need to be replaced. The question is, what other areas need work? Each item that you restore makes the adjoining parts look even worse.

ING YOUR PROJECT

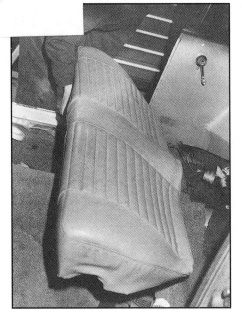

To really get an accurate picture of the interior's condition it helps to remove the seats. Under the back seat cushion you will usually find a mini garbage dump. A comparison of the hidden upholstery material and the exposed surfaces will show the degree of wear and fading.

A thorough inspection of the interior components will save you from future problems. Look closely at the seat springs for breaks or rust damage. A seat that sags probably has spring damage.

Many upholstery jobs can be done by professional upholstery shops or in a home shop. This classic Chevy interior was restored in a home garage. The pros use the same reproduction upholstery kits as you do. They can get the job done much quicker, but they charge a lot for their time. You can save a considerable sum by doing the work yourself.

Chevy pickup only to discover that the truck wasn't driveable with the new seat. I was disappointed, but the pros at the upholstery shop were out a considerable amount of time and money since they had to start over.

A six-way power bench seat out of a Lincoln sedan was purchased at a wrecking yard. The width was perfect and the price was reasonable. The shop used real mohair upholstery with hand-stuffed pleats. The seams were perfect; the pleats even matched up whether the fold-down arm rest was up or down. The seat was very comfortable. It was a work of sewing art.

The seat looked great in the truck. I eagerly got in and promptly jammed my knees into the dashboard. No amount of adjusting the power controls would yield enough leg room for me to operate the truck. The problem was the angle of the seatback. The upholsterers failed to measure or account for the substantial difference between the relatively upright pickup seatback and the more reclined sedan seatback. The seat was a fixed bench seat so it couldn't be adjusted. The internal framework would have to have been custom built to fit the truck, so the shop chose to start over with a properly measured seat.

The focus of this book is how to do easy automotive upholstery jobs in your home garage. That limitation means that you will occasionally encounter jobs that are beyond the scope of home upholstery. With experience and the right equipment you could learn to do these tasks, but in the beginning it is good to understand your limitations. By being realistic about what you can and can't do, you run very little chance of damaging your car.

DO-IT-YOURSELF vs. PROFESSIONAL TASKS

The more complex the task, the more you should consider getting professional help. If you want a one-of-a-kind custom interior, you will need to find a talented pro. If you need to repair a complex item like a molded, padded dashboard, these jobs usually require expensive machinery that are beyond a home shop. You could patch up a minor tear in a

PLANNING YOUR PROJECT

padded dashboard or replace a worn pad if reproduction ones exist, but to totally restore a wasted pad, you need to send it to a professional.

If your car has exotic or expensive materials like leather, chances are that you will need the help of a professional. Most of the upholstery kits and related restoration products are geared toward less expensive (and easier to work with) materials like vinyl. If you have a rare or expensive car, you might not want to chance devaluing it by making it your first upholstery project.

Even if you aren't overflowing with confidence about your upholstery skills, it is possible to save money by doing part of the work yourself. You can easily do the preliminary tasks such as stripping and cleaning the interior. You can also reassemble the interior once the professional has done the recover work.

If you want to upgrade the material quality over what is readily available in kits, you will need to seek out a professional upholstery shop. Custom carpeting is a good example of this. There are excellent factory molded carpet kits, but if you want Mercedes quality carpet in your Volkswagen, you will need a pro to cut, fit, and sew a custom carpet.

I'm not trying to discourage you from doing your own work. I only want you to be realistic so that you will be pleased with your results rather than disappointed. As long as you stick with the more basic fabrics and rely on the excellent upholstery kits, you should be able to do a tremendous amount of work yourself.

Upholstery Kits

The key word separating do-it-yourself tasks from professional jobs is *sewing*. Until you can master the operation of an industrial sewing machine, stick with the jobs that already have the sewing done for you. This means using pre-sewn upholstery kits. The great news about these kits is that the quality is very high for most of them, they are reasonably priced (volume production equals consumer savings), and the range of cars and trucks with upholstery kits is expanding constantly.

Popular cars like Mustangs, Falcons, Camaros, Chevelles, Novas, Corvettes, Volkswagens, Chevy trucks, and many General Motors muscle cars are quite well served by a number of reproduction upholstery companies. Owners of popular Chrysler products are finding more and more kit availability all the time. If there is a strong enough demand, sooner or later, someone will step in and service it. The only problem with the more limited demand kits is that prices are usually higher.

Generic Kits—There are even a lot of generic upholstery kits. These kits are some of the best upholstery bargains around. Since the upholstery isn't an exact factory match, the companies can offer these kits at very attractive prices. These kits are available in a wide variety of materials including vinyl, cloth, tweed, plain designs, and custom designs. Generic kits are great for daily driver cars and trucks or a

The wide array of upholstery kits is what makes it so easy to upholster your car or truck at home. The people who make the kits usually specialize in a particular brand of car, so they are experts at getting the upholstery perfect. They do all the difficult work for you.

High quality upholstery kit manufacturers do the necessary research in order to make their kits correct. They check factory fabric samples, original factory photos, and pristine original cars in order to make sure the kits are exact reproductions.

Planning Your Project

Upholstery kits are first made in the most popular colors such as black, white, and red. The color selections are expanded as the demand warrants. Some companies (like C.A.R.S. in Berkley, Michigan) have extensive supplies of original fabrics so they can often sew you a set of seat covers in a color that isn't currently reproduced.

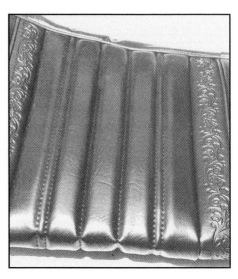

As good as the overall quality of reproduction upholstery kits is, you still need to shop carefully. Some of the earlier kits scrimped on details in order to keep costs down and to make manufacturing easier. The quality kits have the correct embossed patterns (where required), the right stitching, and the proper weight of vinyl.

vehicle that you want to sell without sinking a lot of money into a full restoration.

Material Type & Quality—Some companies will proudly advertise that they still have factory-original material, but this isn't always better than the reproduction material. Sometimes the original material used could be quite shop-worn or it might have even been "seconds" when it was originally produced. Then, there are cases where the reproduction upholstery is actually as good as or better than the old stock.

There are situations where no reproduction material exists. This is particularly true with unusual color combinations. In these cases you are forced to use whatever original material you can find. Since tooling up for reproduction upholstery is expensive, restoration companies produce the more popular vehicles and the most popular color and fabric combinations first.

The restoration companies that have been around for a while know their products. They do their research and try hard to stock the most accurate reproductions. These companies do all the difficult research so that you get the most accurate materials possible. You are more likely to find some marginal reproduction fabric or some early reproduction material at a swap meet or from someone who gave up on a project that was started many years ago.

To avoid getting marginal material, we suggest that you deal with well-established, reputable companies. Mail-order companies can offer attractive prices, but make sure that you understand their return policies in case you don't like the upholstery material. If you think you might have any doubts, buy your upholstery kits from a local retail restoration supply store. This way you can inspect the material before you purchase it. We used both local and mail-order suppliers for the illustrations in this book. We had excellent results from both sources, but we once got the wrong year seat back panels. Fortunately, we bought the panels at a local Mustang restoration source and it was a quick and easy matter to exchange them for the correct panels.

To give you a little better idea of what you can and can't do yourself, we will highlight some of the most popular areas of automotive upholstery.

Carpet

Replacing carpet is one of the easiest (if not the easiest) automotive upholstery tasks. The reason it is so easy is that there is a tremendous amount of pre-cut, factory molded replacement carpet available. Most carpet jobs are simple replacement operations.

Potential problems arise when you have a unique vehicle that requires custom carpeting. If the original carpet isn't all ripped up, it can be used as a pattern for the new carpet.

Planning Your Project

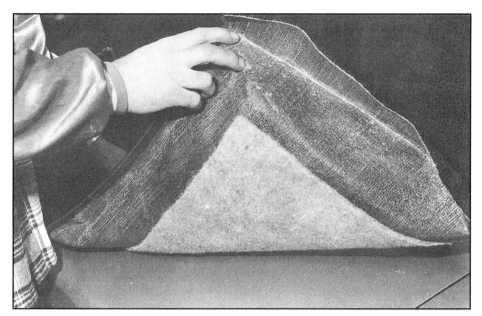

You can buy plain carpeting that can be used for cars, but it is difficult to make it look right. The carpet kits are usually molded so that they conform to the unique shapes of your vehicle's floorboard. The molded carpets usually have the jute padding attached to the bottom of the carpet.

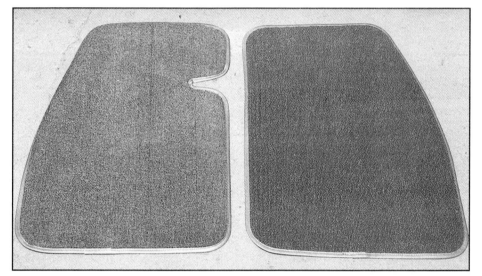

It is also possible to get matching floor mats for most carpet kits. These floor mats are a great way to keep the carpet in good condition.

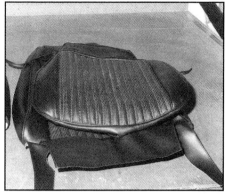

Seat cover kits usually come in four pieces for bucket seats and three sections for split bench seats. Some bucket seats use a separate seat back panel. The back panels can be either upholstered or colored plastic units. Always check the seat covers for any flaws before you start punching hog ring holes in the material.

Since it doesn't make economic sense to mold a one-off carpet, the variations in the pattern need to be cut and sewn. Carpeting is heavy material so you will need an industrial sewing machine.

If you want a non-standard type of carpet or if you have a vehicle (like an old truck) that didn't originally come with carpeting, you will probably need professional help. It is possible to buy the carpet material, make the patterns, cut the carpet to size, and glue the pieces to the floor. If done well, it will probably look OK, but not as seamless as a professionally sewn version. When you glue a carpet to the floor, it is tougher to make any repairs or to remove the carpet for any maintenance work or cleaning.

Overall, carpeting is an area anyone should be able to handle as long as you stick with the basics.

Seats

Seats, whether bench or buckets, are the most frequently repaired, reupholstered or recovered area in automotive upholstery. Seats are most likely to wear, the most visible, and the largest component of any interior. As mentioned above in the discussion about upholstery kits, there is a large selection of available seat kits. As long as you use a kit, you should be able to reupholster or recover your vehicle's seats.

In rare cases, you may find a vehicle where only the front seat kits are offered. You might need a pro to make matching rear covers. Of course, front seats get the most wear unless you own a taxi cab. The vast majority of upholstery seat kits are made for two-door vehicles because most people save, restore, and/or modify two-door vehicles. If you have a four-door sedan or a station wagon, you may have trouble finding rear

Planning Your Project

You can find good used seats at many wrecking yards. If you are using seats from a different brand vehicle, take plenty of measurements before you buy the seats. These Mustang deluxe "Pony" seats were for sale at the Wild West Mustang Ranch in Woodinville, Washington. With seats like these, you can simply bolt them in and be on your way.

Back seats are usually in much better shape than the front ones. If your back seat is really nice (or if you just want to save some money), you can buy the front seat covers by themselves. The price is usually more attractive when you buy both the front and rear seat covers.

cost of this work, it may or may not be less expensive than recovering the whole seat with a kit.

If you would like custom materials or colors for your seats, you should see a professional. A professional can add custom touches like storage pockets on the back of bucket seats. If you want extra lumbar support, a pro can build a custom seat base or add special padding.

Many people like to install seats from a later model luxury car into an old truck or car. If the seat is an exact fit, you can do the job yourself. If the seat requires narrowing (often the case with old trucks) you can narrow the seat yourself, but you will need a pro to section the upholstery and sew it back together. If you want to match the luxury seat upholstery to your stock back seat or the other way around, you will need professional help.

The key to doing your own seats is to use upholstery kits whenever possible.

Door Panels

If someone makes reproduction door panels for your vehicle, they are simple remove-and-replace parts. The job becomes much more difficult if you need custom panels or if your car has molded panels (like padded dashpads). Some companies can restore molded door panels, but you usually have to send away for this service.

Some upholstery kits are available with custom fabrics. These kits usually come with matching door panels. These kits are most common for classic Chevys, street rods, VWs, and some trucks.

Making custom door panels is one of the easier "advanced" upholstery

seat kits.

If you have seats that are in good condition except for a single rip, it would make sense for you to remove the cover and take it to an upholstery shop to be mended. Upholstery shops can install patch panels in upholstery where just the main seating surface is worn and the side and back sections are still serviceable. Depending on the

Planning Your Project

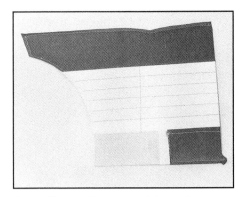

Depending on the age and type of car you have, the side and door panels may be upholstered, molded plastic, metal, or a combination of styles. Upholstered panels like this rear seat side panel are already sewn to the backing board. You need to cut out any window crank holes.

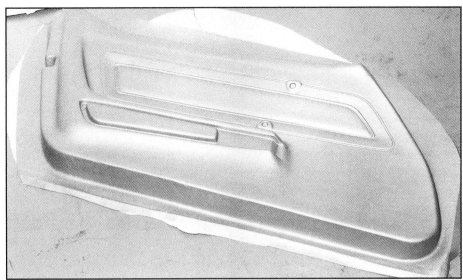

Injection molded plastic reproduction door panels look like this when they first emerge from the mold. The factory trims the excess material, but they usually leave the door handle openings for you to cut with an X-Acto knife.

projects since the panels are basically flat. If you can accept a simple design, door panels can be made with glue instead of sewing.

Like car seats, door panels with custom features like map pockets will probably be best left to the professionals. It is fairly common to find door panels that are quite worn along the bottom edge. If the panels have a natural styling break a few inches up from the bottom, it is possible to have a shop just replace the bottom section. Determine if it would be cheaper to buy a complete reproduction door panel rather than repair your existing panel.

Headliners

Headliners are the classic good news/bad news scenario. The good news is that there are lots of pre-sewn, high quality kits available and that headliners can be easy to install. The bad news is that headliners are easy to damage and they can be a bear to install. A poorly installed headliner is a sure sign of an amateur upholstery job.

The two biggest problems with headliners are getting them nice and taut and working in the most awkward part of your interior. Professionals know all the tricks to getting a drum-tight headliner. We will show you several techniques to get a good installation, but this is one of those jobs that gets easier with experience. Headliner material is often rather thin and relatively easy to damage.

Unless you really botch a headliner installation and have waves of sagging material, most people won't notice minor flaws. Black headliners are best for first-timers because they hide flaws better.

Working almost upside-down can take a little getting used to, but it is easier if the seats are removed. It is also difficult to install the edges of some headliners where they meet the rear window. Some headliners are meant to be installed with the rear window and/or windshield removed. If you don't feel confident about your ability to handle expensive glass, you might want a professional to do the job.

Dashpads

The toughest part about replacing dashpads can be gaining access to all the retainers. You often need to work right up against the windshield, which can be difficult. If reproduction pads are available, the job is mostly a remove and replace operation. If you have a rare vehicle or one with no current replacement pads, you will need the services of a professional restoration expert.

There are kits to bond and repair cracked dashpads. The results can be mixed. This type of repair might be fine on a transportation car, but could be a real eyesore on a restored vehicle. A detail shop that specializes in inexpensive repairs for used car lots might be able to help you with a simple rip or crack in your dashpad.

If you want a dashpad for a vehicle that didn't come with one from the factory or if you want a special material, you will need professional assistance. Sometimes dashpads just get faded from the sun. You can probably respray the pad if it isn't

Planning Your Project

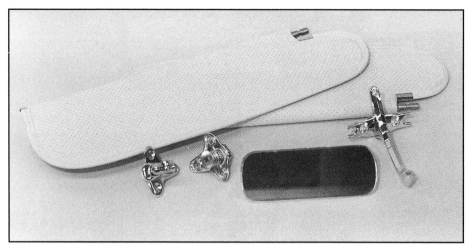

When you install a new headliner it will make the old sun visors look shabby by comparison. Fortunately, there are lots of reproduction visors for popular cars. You can also get reproduction mounting brackets and rear view mirrors.

There is quite a lot of basic disassembly and assembly work in restoring an interior. Even if you are going to farm out some of the work to a professional, you can still save as much as $60 per hour by disassembling the interior yourself.

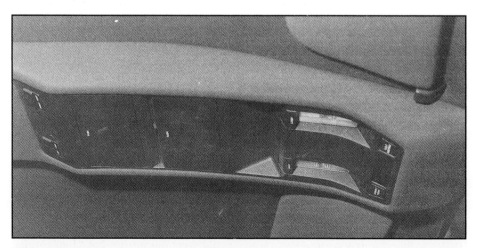

Many newer vehicles (especially mini vans) have overhead consoles with storage compartments and reading lamps. The only thing to watch out for is the contour of the console versus the contour of your roof. You might want to bring along a cardboard template of your roof contour when you are looking around at wrecking yards.

glued to the dashboard. Glued pads are very difficult to remove without virtually destroying them. It is possible to respray a pad that is still attached to the dashboard as long as you carefully mask off the rest of the interior.

Color Changes

If you want to simply change the color of an interior that is in good condition, you can do that yourself. Be warned that long-lasting color changes can be difficult to achieve. Even pros can have trouble here because the material will soon get worn. Unless you can dye the material, most color changes only affect the surface of the material. That is why they are so easy to scratch and fade.

The difficulty of a color change also depends on what color you are applying and the darkness or lightness of the original color. Some bright colors (like red or orange) are very difficult to color. The general rule is to cover a lighter color with a darker one, preferably black.

A professional will probably use better quality commercial products than what you can get in most auto supply stores. He will also be more experienced at properly cleaning the original materials, which is a key element of a successful color change. You can learn these techniques and obtain the materials, but understand that good color changes require more work than a little spray can magic.

Trunks

The major extent of reproduction materials for trunks is simple trunk mats. These mats are simple remove and replace products. If you want a carpeted trunk, you will need some

Planning Your Project

Don't pay big bucks for a new faactory sun visor unless you have to. Wrecking yards are an excellent source. If you're going custom, many of the newer visors with mirrors and lights can be adapted to older cars and trucks.

When you do your own work, you can afford to take extra time for the little details. Painting the front seat tracks with black rust preventative paint is a case of over-restoration, but they look better (and aren't as obvious) than the unpainted tracks.

custom sewing help from a professional. It is possible to cut some flat carpet for the main floor of the trunk yourself, but if you want seamed edges like in a luxury car, you will need a professional sewing machine. You could cut your own carpet and have the pieced hemmed by a professional. This is also known as the *binding* on a carpet.

A trunk is a good place to experiment. Even if you aren't super confident of your abilities, the trunk is a safe place to learn. Chances are that you will do a good job, but if you don't, how often do you look in the trunk?

Other Areas

There are many other aspects to restoring an interior, including sun visors, package trays, center consoles, dashboards, steering wheels, aftermarket seats, glove boxes, seat belts, and chrome/stainless trim restoration. As with the above projects, much of it can be done by the first time do-it-yourselfer. It is only the rare cases where professional help will be required. Study the rest of this book and you will have an even better idea of what you can accomplish at home.

BEFORE OR AFTER PAINT

One of the most common planning concerns is whether to do the upholstery work before or after the paint and bodywork are performed. Sometimes other factors (e.g. money, shop availability, weather conditions, etc.) dictate the sequence of work, but if you can plan ahead, it usually works best to do all the paint and bodywork before tackling any upholstery work.

A problem that can crop up is not knowing how far you want to go with a particular vehicle. If the paint is okay and the interior is a mess, you will obviously be interested in fixing the interior. No one likes to drive around in a torn-up interior. The problem gets worse after you fix the interior and now the previously acceptable exterior starts to look less acceptable next to the revitalized interior. Yet, at the start of the project you didn't feel like restoring the entire car.

When to paint is a tough call. The more you like the car, the more you will want it to approach perfection. There is also a lot to be said for not spending as much as $5,000 on paint and bodywork and just driving the car. The best news about deciding to improve the paint and bodywork later is that you can remove the interior and reinstall it after the messy work is done. This is more work than doing the bodywork first, but you won't waste a lot of money.

Painting Before

But even if you are very careful, it is best to do the paint and bodywork before the interior restoration. You can save a lot of time because you don't have to worry about perfectly sealing off the interior. Paint and bodywork generate a lot of dust and

Planning Your Project

This beautiful custom mohair upholstery was fitted to the stock '57 Chevy bench seat by Gary Cler of Trimcraft Custom Upholstery in Snohomish, Washington. Craftsmanship like this requires considerable talent and experience.

overspray. Overspray can be particularly difficult to remove from the interior.

Another good aspect to painting first is that you can paint any exposed metal interior surfaces. You can even paint a couple of inches inside the edges of door panels and similar parts. This way, if there are any gaps in the upholstery, you will see the same color as the exterior. This type of thorough painting makes it very difficult to tell if a car has been repainted.

I have experienced both failures and successes with painting the car after the interior was finished. I had a customized van with a trick interior consisting of teak paneling, custom cabinets, high performance seats, and a lot of plush, black carpeting. Due to scheduling conflicts, the interior work was done after the bodywork, but before the paint work. The paint shop was very experienced and removed the seats. They masked off the interior with big sheets of plastic. They applied several coats of super bright Corvette Yellow enamel. When the interior was unwrapped, there was a dusting of yellow all over the black carpeting. There must have been a leak in the plastic. It took a long time and a lot of effort to clean the interior, but it never was as nice as if the vehicle had been painted first.

Painting After

On the positive side, one of the vehicles that appears in many of the photos in this book was also painted after the interior work was finished. This vehicle was owned by a student with limited resources. He was able to afford the interior long before he could save up for a custom paint job. He enjoyed the restored interior for over a year before painting the car. Extreme care was taken with the masking and the interior was partially disassembled. The result was a great paint job without damaging the interior.

If you know what color you want to paint your vehicle, or if you plan to repaint it the existing color, you can lessen the problems associated with painting after the upholstery work has been done. Many people remove the interior and paint the door jambs and any exposed metal interior parts. Then they do the interior restoration. Before the exterior is painted, the interior is sealed off and the doors are back-taped. The vehicle is painted without ever opening the doors. The chance of paint getting on the upholstery is minimal. The only downside is a slight ridge where the doors were back-taped, but this ridge can usually be rubbed out.

STOCK VS. CUSTOM

Another planning concern is whether to restore the interior to one hundred percent stock or to add custom touches. This is largely a personal matter, but it can have financial ramifications. It is hard to go wrong with an authentically restored interior, but many older vehicles were nowhere near as comfortable as modern cars.

A stock interior will work well in either a stock vehicle or a modified one, but a modified interior will turn an otherwise stock car into a modified one. Many collectible cars are worth more money in all-original condition, but some are just as valuable in a modified state. You need to know the market for your particular car or truck.

The most important factor is your personal taste. If you build a car to suit yourself, it is hard to go wrong. Unless you are strictly in it for the money, it doesn't make much sense to build an uncomfortable car. The downside of building a unique vehicle is that if you ever decide to sell, a personalized vehicle is much harder to sell than a stock one.

In some cases (particularly with less popular makes and models) your choices are limited by what is available. There might not be a big enough demand for reproduction upholstery kits for your favorite car. Or, a common situation is that the reproduction upholstery is only available in limited colors.

Black is the most commonly available color in reproduction

PLANNING YOUR PROJECT

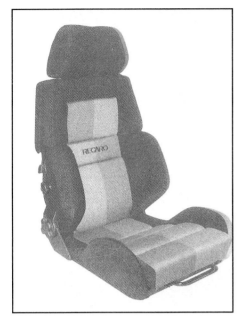

The easiest way to get a custom interior is with aftermarket seats like these trick Recaro bucket seats. The sewing is already done for you. Many custom seat companies will install special fabrics or match the seats to the rest of your interior for an additional charge. You can also get extra matching material if you want the rear seat covered.

Some types of custom door panels can be made without sewing although these panels were custom sewn by Gary Cler. Very thin wood veneer or blank panel board is used for the base. The kind of custom panels that are easiest to make involve gluing tweed fabric over a thin layer of foam.

upholstery kits and it is hard to go wrong with a black interior. Even if your interior was originally gold or some other less popular color, it will probably look fine in black. The only problem is that the color won't match the data tags and this could bother a picky buyer.

A totally customized interior is expensive and there isn't too much you can do yourself. A good compromise is to install a few bolt-on custom items like seats and a steering wheel. Items that bolt-on are good because the stock interior can easily be put back.

I have built several vehicles where custom, orthopedic bucket seats and a comfortable leather steering wheel were installed. These items greatly add to your daily driving comfort, yet they can quickly and easily be removed when you sell the vehicle. I have even had cars where I restored the factory bucket seats and set them aside until it was time to sell the car. I drove the car with aftermarket bucket seats and saved them for my next project car. If you stick with basically black (or the same color regardless of what it is) interiors, you can do what I have done and transfer the expensive orthopedic seats from one car to another.

Areas where you can blend stock and modified interiors is with carpeting and insulation. Most carpet kits are factory duplicates, but if you can find a better grade of carpet, go ahead and install it. Few people will view premium carpet as a negative and if they do, the factory style carpeting kits are inexpensive and easy to install. You could also add extra insulation and sound deadening material without anyone noticing. If they did notice, who is going to complain about a car being too quiet?

The biggest thing to consider very carefully in the stock versus modified area is any change that requires cutting up sheet metal. A cut-up dashboard (common for custom sound systems) or holes in door panels can be difficult to fix later. Try to find a way to accomplish your goals without butchering the sheet metal. If you need to make modifications to the floorpan to install special seats, remember that once you make major modifications, it is increasingly difficult to return the car to stock. Modified vehicles are a lot of fun, but understand the pros and cons before you cut into a pristine car or truck.

PARTIAL OR COMPLETE?

I recommend doing the entire interior restoration at one time whenever possible. This concept ties in with removing all of the interior and doing the paint work first. This is the most logical way to completely restore a car or truck.

Planning Your Project

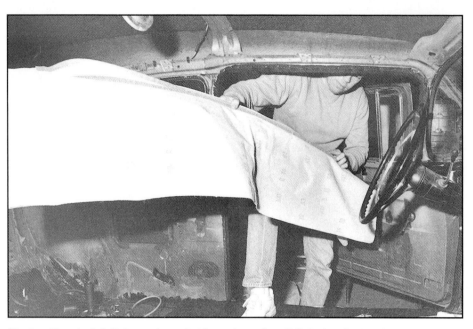

The headliner is definitely one item that is easier to install if the interior has been completely stripped.

The dilemmas come when you're not painting the car or when only part of the interior needs repair. It is perfectly understandable that someone would replace torn seat covers; the perennial question is "Should such and such be replaced while we're at it?" You can build a whole car on the "while you're at it" program.

Pros & Cons

The main concerns of partial versus complete interior projects are making the job as easy as possible, not damaging previously restored areas, the ability to use the vehicle, and saving time and money. Some interior work can be done out of the car and some must be done inside. Depending on your vehicle, the inside work can involve some contortion skills. For that reason it is nice to have as much room as possible.

When the car is gutted, you can restore the interior in a logical progression. The headliner is easiest to replace when the seats are out of the way. Another benefit to having the seats out is that you are not climbing over them and dragging equipment in and out of the car. This lessens the chance of inadvertently damaging the seats.

The less you take parts in and out, the less chance there is for damage. It makes sense to do any carpet work when you have the seats out for recovering. The "while you're at it" factor is any rust in the floorboards. Any entrance points for moisture will only damage your new carpet so you might as well do those repairs "while you're at it."

Most people want to get projects completed as quickly as possible. This is normal, but remember that it is false time economy to recover your seats and then have to go back later to fix a broken spring or add more foam to the seat buns. The more you can do the first time, the shorter the total time investment will be.

The two best cases for doing partial interior renovations are lack of funds and the need to keep driving the vehicle. When finances are a problem, there is much you can do. If possible, consider doing items like the carpet first. Even though you will need to remove the front seats again when they are recovered, you won't have to bother with items like the door sills, seats belts, or center console.

A tactic for keeping your car on the road as much as possible is to do the big jobs on a weekend and to break the tasks into small units. For example, with bucket seats, recover one at a time. You only need a driver's seat to use the car. The type of work that will most disable your car is anything that requires removal of the steering column, windshield or rear window. This is necessary on some cars for dashpad work or headliner replacement. Plan for the disabling tasks when you least need the car and you should experience a minimum amount of down time.

You can't go too wrong regardless of what sequence you choose for your interior restoration. Planning will make the job as easy and pleasant as possible. Take a little extra time for planning and you will be glad you did. ■

Tools & Supplies 2

A new project is usually charged with enthusiasm. The urge to start tearing things apart is strong. However, temper your enthusiasm with a little patience. Blindly rushing into a project can lead to costly mistakes if you haven't done your homework.

By homework, I mean being fully prepared before you disassemble your interior. Have your tools, supplies and materials on hand. Have a suitable work place. Study any pertinent reference material. Double check your upholstery kit to make sure it is complete, and that it is the correct one for your car. Most companies will gladly correct their own mistakes, but they won't necessarily make good on yours, so don't start installing hog rings until you are sure that you have the right kit for your application.

One of the best things about basic automotive upholstery is the small amount of tools required to do professional quality work. Unlike so many other tasks that require expensive, sophisticated tools, most upholstery tools are very simple and affordable. The only exception is an industrial walking foot sewing machine, but because upholstery jobs that require custom sewing are beyond the scope of this book, you won't need one anyway.

In other endeavors, like engine building or paint and bodywork, there are big differences between professional grade tools and hobbyist equipment. Look at the differences in price and quality between tools in a Snap-On catalog and similar tools at a discount retailer.

But with automotive upholstery, beginners and professionals use

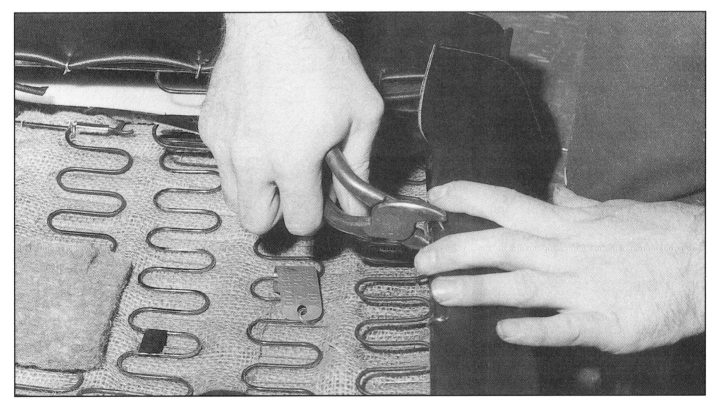

The primary tool used in the installation of automotive upholstery kits is the hog ring pliers. Most upholsterers seem to prefer the bent-nose hog ring pliers. The hog ring pliers push the two sharp ends of the hog ring through the material as it is stretched over the seat's framework. As the pliers' jaws are closed, the hog ring ends meet to form a closed metal loop.

Tools & Supplies

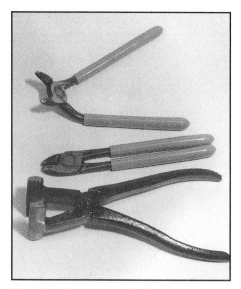

Hog ring pliers are available in either bent (top) or straight (middle) configurations. The angled pliers are spring loaded to stay open, while the straight pliers are spring loaded to stay in the closed position. Which type you use is a matter of preference. The lower tool is a stretching pliers.

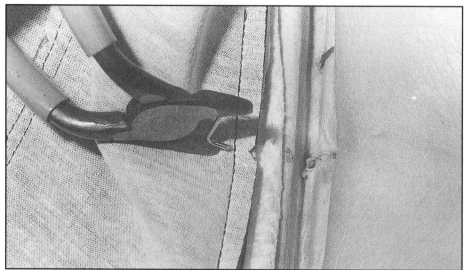

Hog ring pliers are notched so that they hold the hog ring while you position the ring and material over the frame or listing wire. Sometimes you need to pull very hard on the material so it helps that you don't have to worry about the hog ring slipping out of the pliers.

basically the same tools. Professional upholstery shops may have special machines for making fabric-covered buttons, high-speed fabric and foam cutters, steamers for removing wrinkles, and pressure-pot glue guns for applying a lot of contact cement in a hurry, but making buttons isn't necessary with upholstery kits and the other tools can easily be supplanted with inexpensive alternatives.

A tool that professionals pay much more attention to than amateurs is their scissors or shears. They are very protective of their personal favorites and keep them clean and sharp. The pros need to cut or "trim" a lot of raw material, since they fabricate most of their work. This is where the term "trimmer" comes from at upholstery shops. You won't need scissors much if you stick to upholstery kits. The little cutting that may be required can be accomplished with an ordinary pair of household scissors. This is an example of how basic the required tools are, and how upholstery kits eliminate much of the hard work for you.

The supplies needed for basic automotive upholstery are as simple and inexpensive as the tools. By supplies, we mean the gluing, fastening, cleaning, and painting products as opposed to the materials (fabrics and carpeting) that make up the elements of upholstery and carpeting kits.

You won't need all the tools or supplies mentioned in this chapter for every upholstery job. Study the chapters that deal with your anticipated tasks and just buy the tools as you need them. The following is an overview of the basic automotive tools and supplies.

TOOLS

Most of the tools needed for basic automotive upholstery tasks can be found in the average homeowner's tool box. There are very few specialized upholstery tools required and they are easy to find and inexpensive. The most important specialized tool is a pair of hog ring pliers and people have even been known to get by without them.

Hog Ring Pliers

Hog ring pliers are the basic tool of upholstery work. The special pliers have slotted jaws that hold the hog ring in place while you push the sharp ends of the hog ring through the material to secure it to the listing or seat frame. After the hog ring is in position, you squeeze the hog ring pliers to close the loop.

Hog ring pliers come with different style noses. The two most common styles are *straight and bent*. The bent-nose pliers have about a 45-degree angle to the nose. The pliers are spring loaded, but some are designed to hold the jaws open while others hold the jaws closed. Each of these different designs have their advantages.

I prefer the bent-nose style because they seem easier to push into tight spots. Other people prefer the straight-nose versions. Hog ring pliers are

Tools & Supplies

A couple different types of pliers can be used for various upholstery and interior restoration tasks. Diagonal cutters are used to remove old hog rings. The hog ring can either be cut or twisted out.

very reasonably priced, so it won't break the bank to buy more than one style. Look for pliers with vinyl handle grips. They are easier on your hands when you are installing a lot of hog rings.

I have seen people use hose clamp pliers and even common pliers to install hog rings. The slotted nose on the hose clamp pliers are fine for gripping the hog rings, but they just don't compress the hog rings as easily as hog ring pliers. Regular pliers don't hold the hog ring very well and it is frustrating to constantly drop the rings. Also, the noses of these pliers aren't as well suited as hog ring pliers for getting into tight places. Common pliers and hose clamp pliers should only be considered as emergency hog ring tools.

Diagonal Cutters

Diagonal cutters, which are sometimes known as *side cutters* or *dikes*, are the basic tool for removing old upholstery. They cut and remove the old hog rings. They can also be used to remove staples or trim tacks.

Diagonal cutters come in different sizes and nose designs, but the standard 7-inch, bowed handle cutter will work fine. There are smaller styles and special high leverage cutters if you have specialized needs.

Look for cutters with vinyl grips to make it easier on your hands. It is also good to buy quality cutters that open and close with one hand. They will make your work quicker and easier.

You can use common pliers to twist off old hog rings, but that is a lot of work. Diagonal pliers can be used to either cut the hog ring or twist it off. Many upholsterers develop a combination twisting and cutting motion that quickly dispenses with the old fasteners.

Mechanic's Tools

A basic set of mechanic's hand tools is necessary to disassemble and reassemble your interior components. The most commonly needed tools include wrenches, sockets, and screwdrivers. Depending on the make and year of your vehicle, you may need standard or metric tools.

Some interior items are secured with specialty fasteners like Torx (those star-type fasteners that are commonly used on seat belt anchors) bolts and screws, Allen head fasteners (sometimes metric), and even square head screws. If your vehicle has some specialized fasteners, there are usually only a couple different sizes. You

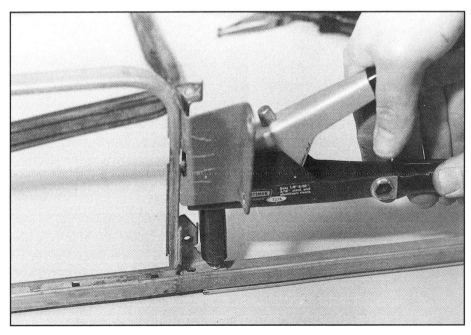

Some components such as this Chevy vent window need to be taken apart for refurbishing. Rivets and a rivet gun are used to put the window frame back together.

15

Tools & Supplies

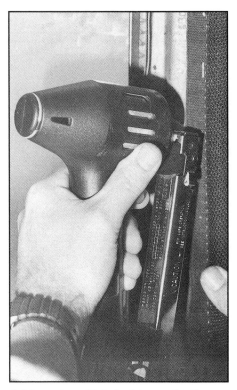

A fair number of interior jobs require the use of a stapler. The professionals use pneumatic staple guns, but a modestly priced electric stapler should be sufficient for restoring a single vehicle. Here, the staple gun is being used to attach the wind lacing to the tack strip around the door opening of a classic Chevy.

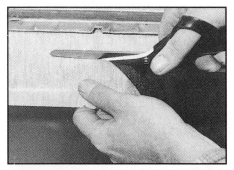

A good pair of scissors (some pros call them shears) are necessary for a variety of upholstery jobs. Items like headliners are made oversized so they need to be trimmed.

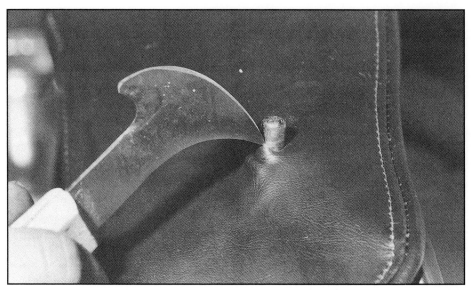

A variety of cutting tools are needed for most upholstery jobs. A good pair of sharp scissors is needed for trimming excess material. Openings need to be made for things like seat pivot pins. A common single-edge razor blade will work or you can use this versatile cutting tool known as a trimmer's knife.

probably don't need to buy a complete set of specialty tools, although they are less expensive when purchased in sets rather than individually.

You may have an occasional need for an electric drill. A 3/8-inch drill should be fine. If your car has a large number of trim screws, a cordless drill or a compact cordless screwdriver can save you time and hand fatigue.

Cutting & Poking Tools

Although the vast majority of all cutting tasks are handled by the companies who make the upholstery kits, you will have an occasional need for cutting materials or poking holes. Professional trimmers rely heavily on high quality scissors. The pros often refer to scissors as *shears*. You need sharp scissors (not ones your kids use to open frozen juice treats), but you don't need the high quality shears used by the pros. All-purpose shop shears, like those sold at hardware stores, will work fine. Your most common cutting situation will probably be trimming carpeting for things like floor shifters or center consoles.

Sometimes you will need to poke holes to locate fasteners or seat pivot points. A scratch awl or an ice pick will handle those chores. You don't want to cut holes that are too big or rip the material.

A utility knife or some single-edge razor blades can be used for miscellaneous cutting tasks. Utility knives work well for cutting the tough backing on carpeting.

Knives—Professionals use special power foam saws to make custom seat cushions or repair factory cushions. You won't have enough occasions to warrant such a specialized tool so some alternatives will do. An electric carving knife will work, as will a sharp hacksaw blade. Various kitchen knives, such as steak knives, serrated bread knives, and those serrated knives used to separate frozen food will also work. Even single-edge razor blades will cut foam. The thing you need to avoid is tearing or ripping the foam when you cut it. Practice on a scrap of foam before you cut into anything important.

There is a wicked-looking knife known as a *trimmer's knife*. This very sharp knife has two curved edges.

TOOLS & SUPPLIES

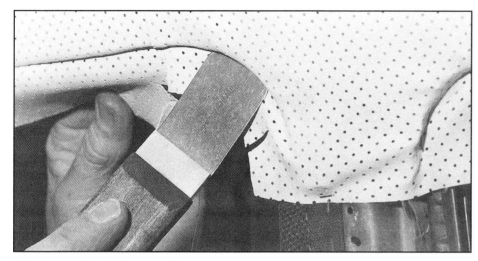

This is a special upholstery tucking tool. It looks like a putty knife with rounded edges. The blade is thick and slightly bent. A regular putty knife can cut the upholstery. Here, the tucking tool is being used to push the edge of a headliner up inside the door frame.

Because the tucking tool is thick, it is strong enough for a variety of prying jobs. Here, the blade is being used to remove old window whiskers from the door.

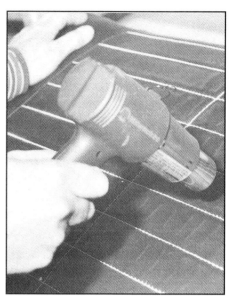

Wrinkles are an unwanted component of most upholstery kits and headliners. The careful application of heat from a heat gun can help remove those wrinkles. A standard hair dryer will also work.

One edge is a couple inches long and gracefully curved while the other edge is a short hook. The trimmer's knife will cut carpeting, weather stripping, and almost anything else. It is also used where seams need to be cut for making patterns out of old upholstery.

Although they are called putty knives, you really don't want a sharp one. A putty knife with a relatively dull tip can be used for prying off items like door panels. Even better than a standard putty knife is a specialized upholstery *tucking tool*. This is basically a modified putty knife with a short, thick blade with rounded corners and a strong wooden handle. The tucking tool can be used for prying and tucking with little fear of damaging the upholstery. You could shorten the blade and round off the corners of an old putty knife to make your own tucking tool.

Heat Guns

Heat guns are good news/bad news tools. They can hurt an upholstery project as much as they can help it. When judiciously applied, heat can remove wrinkles and tighten sagging upholstery, but too much heat will permanently ruin the material. Heat guns are often used to perfect the fit of headliners.

The best heat guns have more than one setting. If you use a gun with only one setting, vary the temperature by holding the gun closer or farther from the material. Besides the damage you can do the upholstery, you can also burn your skin with a heat gun.

Many do-it-yourself upholsterers use an old hair dryer in place of a heat gun. Hair dryers are not as quick, but that can be a good thing for inexperienced users.

Clip Tools

Many cars and trucks use little clips to secure door handles and window cranks. With a lot of effort, you can remove these hidden clips using a flat-bladed screwdriver, but there are inexpensive clip tools that make the job so much easier. With the clip removal tools, you won't run the risk of tearing the door panels.

Tools & Supplies

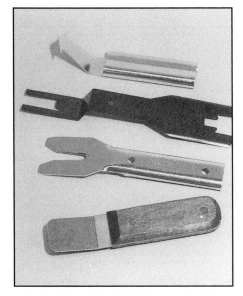

Here are a variety of prying and clip removal tools. From top to bottom: windshield molding clip removal tool; window crank clip removal tool; a door panel clip popping tool; and a tucking tool.

This is a door clip removal tool in action. The prongs of the tool go between the back side of the handle and the nylon washer to engage the little retaining clip. Without this inexpensive tool you would have a difficult time loosening the clip with a screwdriver.

There are a couple different styles of clip removal tools, but the most common one is the slotted version which slides under the door handle and engages the open ends of the clip. You give a little push and the clip snaps off the splined shaft. The less common style clip tool looks more like a pliers.

The clips can lose their tension, but replacements are inexpensive. They also have an annoying tendency to "spring" behind a work bench or some other unseen location. A very specialized tool for closing GM window clips is made by Snap-On Tools.

There are also special clip lifter tools for door panels. Door panels are often secured by plastic or metal clips that are tough to dislodge without damaging the door panel. There are pry-type clip lifters that look like slotted putty knives and squeeze-style clip lifters that look like odd pliers. These clip lifters remove the clip from the door, lessening the danger of ripping the clip out of the upholstered panel. The previously mentioned upholstery tucking tool can also be used to remove door panels.

If you need to remove the windshield or rear glass for headliner replacement, you will need to remove the window molding trim. There is a tool that looks like a strange arrow for removing the hidden molding clips. The tool will free the clips with little possibility of damaging the glass.

Clamping Tools

Occasionally you will need clamps to either secure part of the upholstery or to act as an extra hand. Various sized spring clamps with protective

Spring clamps can be like an extra set of hands on many interior restoration tasks. These clamps are being used to hold a new window channel in place while the glue dries. Clamps are also useful for headliner installations.

Tools & Supplies

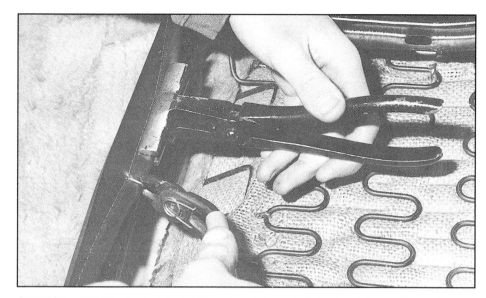

Stretching pliers can be used to grip the edge of the material and pull it over the seat frame. You can hold the vinyl taut while the hog rings are installed. These pliers and many other common upholstery tools and supplies are available by mail from the Eastwood Company in Malvern, Pennsylvania.

vinyl tips work well on headliner installation and even when you are trial fitting a seat cover.

The Eastwood Company offers a special type of clamping tool called *stretching pliers*. These special upholstery pliers have a wide, ribbed jaw that clamps on to the material so that it can be stretched tightly over the fastening rods or seat frames.

Measuring & Marking Tools

If you're using an upholstery kit, you shouldn't have to do much measuring or marking. A standard metal tape measure is handy for checking items or centering things. When you are trial-fitting a seat cover, you may want to measure the distance of seams from corners to ensure a uniform fit.

Chalk—If you need to do any marking, use white chalk. Do not use yellow chalk. Yellow chalk can be difficult to remove. There is a special trimmer's chalk available, but common blackboard chalk, which is softer, will work for limited marking. There is also tailor's chalk, but it has wax in it which makes it difficult to remove. A soft lead pencil can also be used for marking.

It isn't a good idea to mark upholstery with marking pens or ball point pens. You can use them on the back side of carpeting and when marking foam.

Paint Equipment

If you need to change the color of the metal interior parts or change the color of vinyl components, you may find it easier to use a spray gun. Most little jobs can be handled with spray cans, but big jobs will look better when the paint or dye is applied with a spray gun. Large areas tend to look streaked when the products are applied via spray cans. You can get a more uniform pattern and better penetration with a spray gun.

You don't need a big production spray gun for interior work. A small touch-up gun will work well for most jobs. A touch-up gun is also easier for a novice to handle and it is far better suited for tight areas like the bottom edge of a dashboard. Some people even use tiny airbrushes to avoid overspray and lots of taping. The best use for an airbrush is repairing scratches where the original color

Sometimes a little ingenuity will take the place of a special tool. This window channel came as a straight piece. In order to make it curve like the old piece, we bent the channel around a small coffee can.

Tools & Supplies

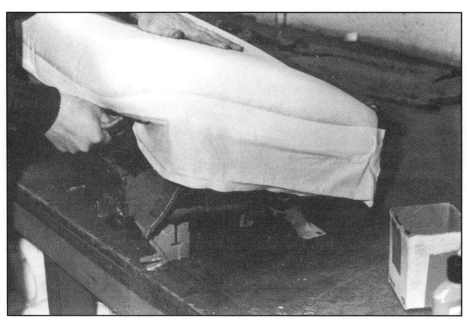

A big work bench or other sturdy, clear space will make seat cover installation much easier. This heavy duty wooden bench was topped with a thin sheet of hard board. That way the top sheet can be replaced if it gets too worn or damaged.

shows through the new paint. With an airbrush you can easily blend the repaired area without worrying about runs.

A one or two horsepower home shop compressor should be adequate for the type of painting encountered in interior work. Big production spray guns require a lot of air, which means you need an industrial capacity compressor. This is another good reason to use a touch-up gun.

Professional shops use spray guns converted to apply large quantities of contact cement. They use a special nozzle that will handle glue. Some shops also use a pressure pot for even larger quantities of glue. Spray glue guns are unnecessary for the type of upholstery projects covered in this book.

Work Area

If you want your upholstery to stay nice, you need a clean, spacious work area. Installing an upholstery kit involves a lot of moving the seats around. You often need to place one part of the seat on the work bench while you wrestle it to get a tight fit at an opposite corner. A greasy work bench will ruin your upholstery.

Even a supposedly clean work bench can damage upholstery. Some clean throw rugs or big pieces of clean cardboard work well as a buffer between the bench and the upholstery.

Good lighting and ample room are important elements of a good work space. You need to be able to move around the car without harming the paint as you remove and replace interior components. Good lighting is very important if you are doing any color change work. You want to be sure that the new color is applied evenly and that there aren't any missed areas.

Upholstery work goes better in a relatively warm environment. Vinyl is difficult to work with if it gets too cold. Any spray painting or vinyl dying also works best in warmer temperatures. The product containers are usually marked as to the optimum temperature ranges.

Reference Material

Information can be your most valuable tool. A factory shop manual has specialized information you need to know about your vehicle. The condensed repair manuals sold in auto parts stores have plenty of good information, but by necessity, these books leave out a lot of material. Repair manuals only present the highlights of the factory shop manuals.

Some shop manuals are divided into mechanical and body related volumes. Information about interiors is usually found in the body manuals. If you need to take apart your dashboard or steering column for a color change, the shop manual will show you all the steps required for doing the job correctly. Not knowing how an intricate little part is fastened to a bigger part can easily result in broken pieces.

If you are dealing with a popular collectible car or truck, there may be books written just for your vehicle. These restoration guides can offer valuable insight into all aspects of the restoration process.

You can provide some reference material of your own if you take notes when you disassemble your interior. Don't trust yourself to remember which way a clip goes or the order in which a component came apart. If you don't want to draw diagrams, consider taking photos with an instant camera or a video camera. There is nothing like a picture to clarify a situation.

SUPPLIES

Most supplies are relatively

TOOLS & SUPPLIES

These Chevy window cranks and door handle and related hardware are all reproduction items. The springs go behind the door panels and the washers protect the upholstery from the handle. The little clips secure the handles to the splined shafts.

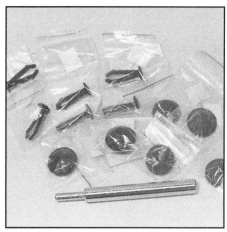

Try to save as many of the original clips and fasteners as possible. The original items will help you be sure that you have the correct reproduction replacement fasteners. You can get most clips through mail order specialty restoration companies or local upholstery supply firms.

common items that can often be found in local hardware or auto supply stores. Specialized adhesives and fabric dyes can usually be found in autobody supply stores and/or large fabric shops if you don't have a specialized automotive upholstery supply store in your area. Many supplies and specialized upholstery tools can be obtained from mail-order companies like The Eastwood Company or J.C. Whitney. Some better stocked fabric shops also carry automotive supplies. Fabric shops usually carry foam and there are specialized foam shops. There are also companies that advertise upholstery supplies in *Hemmings Motor News* under the "Supplies, Tools, Misc." heading. The companies that sell upholstery kits are another good source of specialized tools and supplies.

There are supplies that professional trimmers buy in bulk, but most of these items can also be found in smaller aerosol versions. You will pay more for the convenience of prepackaged supplies, but they will still cost far less than the price of an industrial sized container that would last for dozens of jobs. Even in examples like vinyl dyes where there are bigger color selections in the bulk products, you can spray them with an inexpensive touch-up spray gun which will run on almost any home-sized air compressor. Also, compressors and spray guns can be rented for reasonable rates at tool rental shops.

The basic supply categories include fastening, gluing, cleaning, and painting. Many of the products have alternatives that will work very well.

Hog Rings

A basic supply item of upholstery kits is hog rings. The term "hog ring" applies to the rings that are placed in pigs' snouts to keep them from rooting up the farm. The hog rings used in automotive upholstery are the same design, but usually smaller than even the rings used on little pigs. You can get various-sized hog rings at farm supply stores, but unless you are dealing with some very heavy-duty material, you are better off getting your hog rings from an upholstery supply source.

There are also very small hog rings that are too weak for automotive upholstery. The size that works well for automotive applications is about one inch long on the closed side with a half-inch depth and a half-inch opening. If you have any doubts about which size hog ring to use, ask an upholstery shop which size they use. They may even sell you enough for your project.

Clips & Fasteners

Never throw away any upholstery clip or fastener until the job is complete. Many of the unique little fasteners that are in your car will need to be reused. Even if fasteners or clips are damaged, save them so that you can find the correct replacement items. Upholstery kits usually don't come with any special clips. You can find these parts at car dealer parts departments, but they don't always have fasteners for older models. There

Tools & Supplies

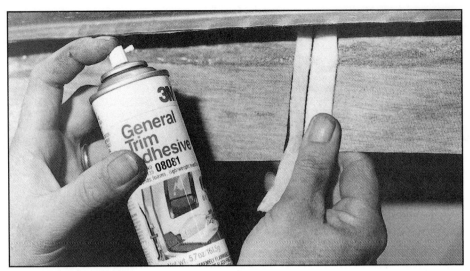

A general trim adhesive such as the ones made by 3M can be used for a variety of tasks. This adhesive is a type of contact cement, so both surfaces need to be sprayed and allowed to dry before they are joined.

are companies that specialize in these parts. Sometimes you can find the fasteners at automotive swap meets, but otherwise, you will need to consult the ads in *Hemmings Motor News*. The best way to avoid problems is to carefully remove your original fasteners and clips so that they can be reused.

Automotive paint and body supply shops often carry some of the more popular clips and fasteners. Sometimes you can find clips that aren't exactly right but are close enough to work.

Two-part Velcro tape can be used to help secure door panels. The Velcro holds well and you can still remove the panels easily. You can glue door panels to the doors, but don't plan on removing them without damage.

Adhesives

Spray can adhesives are the way to go when you are restoring a single vehicle. Top quality automotive adhesives like 3M General Trim Adhesive should always be used. Adhesives are no place to save a few cents. You can usually find 3M products at automotive paint and body supply shops. There are other good brands of adhesives available locally and through mail-order companies. The main thing is to be sure that you get heavy-duty adhesives that are suitable for automotive materials.

If you need to replace any weatherstripping, autobody supply stores carry suitable adhesives. 3M makes a complete line of weatherstripping adhesives. Mail-order companies like The Eastwood Company also stock appropriate adhesives.

Masking tape is adhesive-backed and if you need to repaint any part of your interior, be sure to use automotive quality masking tape. Lightweight household or "bargain" brand tape won't do a good job. Remember to remove the masking tape as soon as you are through painting so you won't have to deal with leftover tape residue.

Cleaning Products

You will probably be surprised just how dirty your interior is when you start restoring it. While you have the interior apart, you might as well give it a thorough cleaning. Many original parts can be reused if they are cleaned and treated with upholstery protectants.

Another time for cleaning products is when any interior parts are repainted or dyed another color. The paint or dye won't adhere properly unless the underlying surface is as clean as possible. You can use a variety of household cleaning products as well as specialized automotive products. Most cleaning needs to be done in stages for optimum effectiveness. Painted surfaces respond well to wax and grease remover. This product is available at automotive paint supply stores. Follow all product directions so that you don't leave any unwanted residue or lint.

A clean surface is especially important when using vinyl dyes. Automotive paint supply stores sell special prep solutions that are designed for the dyes.

There are special brushes designed for cleaning vinyl, but most scrub brushes will work. You want a brush that will clean without damaging the material. Lots of dirt collects around seams and piping, but vigorous scrubbing can also weaken the stitching. Little detailing brushes or even old toothbrushes work well around seams.

When repainting items like metal door panels, dashboards, or metal window moldings, Scotchbrite® pads work well for cleaning and scuffing the old paint. Stainless trim items like door sill plates can be gently cleaned with extra fine steel wool or bronze wool. Bronze wool is a gentle way to clean without scratching.

After stainless or chrome trim items

have been cleaned, they usually need to be polished. A metal polish like Simichrome Polish works well to bring back the luster on dull trim pieces. You will be amazed at how a little elbow grease can restore tarnished trim items.

Paints & Dyes

For the best job, use the best products. With all the work involved in restoring your interior, this isn't the place to save a few dollars with cheap paint or dye. The advantages of bulk paint and dye were mentioned in the tool section, but even if you use aerosol versions, pick quality products.

You can get vinyl dyes at your local auto parts store, but we strongly recommend spending a little more for the products sold at paint and body supply stores. It has been our experience that the professional quality dyes last longer and hold up better than the inexpensive, single step products. Get the cleaning solution and/or surface prep materials that are designed to work with the dyes.

With paint products, be sure to use ones that are designed for interior applications. You want the correct finish. If the factory used a semi-gloss finish, a gloss finish will look out of place. An automotive paint supply store can help you get the right color and right finish. If you have a popular collectible car, you can probably get the proper interior paint from a mail-order restoration source for your vehicle.

Foam

If you are using upholstery kits, you probably won't have any reason to buy bulk upholstery material, but you

When you have the old carpet out, check the condition of the floorboards and the corresponding seams. Seams that are in poor shape can be fixed with an application of new seam sealer. Automotive paint supply stores carry a variety of seam sealer products.

may need a little foam. Chances are good that your original seat buns are compressed and/or damaged. You can restore them to their original size and repair damaged areas with automotive upholstery foam. A common type of automotive foam is Polyfoam®.

You need to obtain foam with the proper density and firmness for your seat. If you are replacing a damaged section of foam, it needs to be the same density as the rest of the cushion. You don't want the lightweight foam sold at craft stores. If you want to add some extra padding for comfort or to compensate for compressed foam, you can use foam that is a little softer.

Depending on its thickness, foam comes in either rolls or blocks. The thinner foam is rolled, while the two-inch-plus foam comes in blocks. Block foam can be cut to suit your repair needs. You can also make thicker foam by gluing two thinner pieces together.

You can get automotive upholstery foam at upholstery supply stores.

Many larger cities even have specialized foam stores. If you are in doubt about what type of foam to use, ask an upholstery shop if they will sell you a small amount.

WHERE TO BUY KITS

Where you buy your supplies and upholstery kits is both a matter of personal preference and practical convenience. Where you live has a great deal to do with your choices.

Clubs

If your car or truck is supported by a national or local club, consider joining the club. Club members are an excellent reference source. They know from past experiences which companies are best and which kits are the most accurate and fit best. Even if you don't belong to a club you can still talk to owners of similar vehicles at local car shows and swap meets. Most automotive hobbyists are very willing to share information with fellow enthusiasts. Some of the larger

Tools & Supplies

There are a variety of sources for upholstery kits. There are many excellent mail order firms and some of them have retail outlets. It is nice to buy an upholstery kit where you can check to make sure it is the right kit. Here, Karen Kennedy of the Wild West Mustang Ranch in Woodinville, Washington, is showing a '65 bucket seat kit to Sean Wallner.

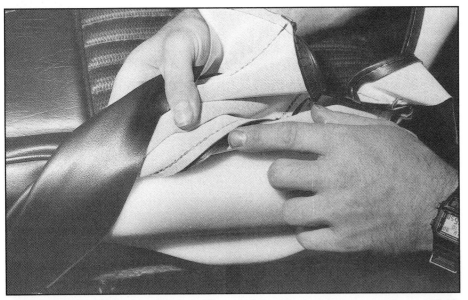

When you buy an upholstery kit in person, besides making sure that you have a '65 kit (instead of a very similar '66 kit) you can also check that all the seams were properly sewn.

car clubs even sell upholstery products to their members direct, often at a good discount.

Mail Order

The farther you live from a big city, the greater the chance you will need to deal with a mail-order restoration company. This is fine as there are many excellent companies that sell first class products through the mail. Unfortunately, there are also a few less reputable companies. I have had excellent luck with mail-order restoration parts companies. I have always dealt with companies that have been around a few years and have good references from friends and acquaintances.

It is best to order the catalog first and carefully check the return policies. Make sure that the company guarantees that each part is shipped in working condition, and that they will accept returns.

Retail

Of course, if you can buy retail, it may be best to do so. This way, you can not only get the parts right away, but you will also be able to inspect the actual product rather than just look at a pretty photograph in a catalog. Depending on the retailer's inventory they might not have your exact kit in stock, but even then, they can usually get items in a day or two. If there is a wholesale distributor in the area, the local shop can often get deliveries by the next business day.

Swap Meets

Many restorers like to buy their upholstery kits and other restoration parts at automotive swap meets. At the bigger meets you can often find displays from major mail-order companies. This is a good way to see the products in person and save shipping charges. Sometimes the companies will have "show specials" to encourage you to buy the parts that day.

Used Kits & Goods—Occasionally you will find upholstery kits and/or used interior parts at automotive swap meets. These items can be very attractively priced because they are often left over from an unfinished project or a project that changed direction. Be wary of upholstery kits

TOOLS & SUPPLIES

It is not uncommon for older cars to have incorrect or missing seats. Swap meets are a good source of used seats, but many seats look very similar so bring along some reference material.

that don't have their packing boxes or an invoice specifying the exact application. Swap meet sellers that aren't completely sure about an application will usually answer your "What does it fit?" question with a "What do you own?" question. Then, amazingly, their parts will be exactly what you need.

A swap meet deal is only a deal if the parts are exactly what you need. A friend of mine bought some bucket seats for his Dodge Super Bee that both he and the vendor thought would work, but they were just different enough not to work. The example seats were from a Dodge Charger. They were the same year seats and my friend only needed the underlying seat buns. He knew that the upholstery wasn't the same, but he already had the reproduction covers (no reproduction seat buns were available at the time). Not only were both cars the same year, the same manufacturer, and the same brand, but they were the same body size series. It seemed logical that they would work, but there were just enough subtle differences in the shape of the seat buns that they wouldn't work with the Super Bee covers. This is an example of how important research and reference materials are for swap meet shopping.

Regardless of where you buy your upholstery kits and parts, inspect the products carefully. At swap meets, do it before you buy. At local retailers, it is a good idea (but not critical) to look for flaws while you are still at the store. If you don't do it at the store, do it soon after you get home. Inspect mail-order parts as soon as they arrive. If there are any signs of abuse to the shipping cartons, have the delivery driver make a notation before you sign for the products. This will make it much easier to substantiate any damage claims.

When you are satisfied that the upholstery kit is complete, correct, and in good condition, put it in a protected place until you are ready to install it. Most kits are pretty tightly packed into their shipping boxes. It is a good idea to unfold the upholstery so that any wrinkles will have a chance to relax. If you can store the upholstery in a warm location, do so. The warmer temperature will help remove any wrinkles. ∎

Painting Interior Parts 3

If the metal parts of your interior are in good condition—not faded, scratched, or covered with surface rust—you can probably avoid painting. If that is the case, then you can read on ahead to other sections of the book to work on individual upholstery and interior areas.

Before you decide to skip any paint work, take a very careful look at the painted parts. The more you improve the surrounding components, the worse the painted parts will look. A window molding that looks presentable in a tattered interior will itself look shabby in the company of new upholstery and carpeting. A non-upholstery example of this phenomenon is how dull old bumpers and chrome trim items look after a car has been repainted.

Most of the paintable interior parts can be removed easily so they can be painted away from the car. The major exceptions are the dashboard and some doors. Many trucks and some cars have only partial upholstery inserts on the doors instead of full, upholstered door panels. These doors are susceptible to scratches and fading. If you are changing the color of your upholstery, these doors should be painted accordingly.

Trucks, especially older ones, require the most work when it comes to painting interior components. Much of the interior is body color. Many trucks only came with headliner inserts instead of full headliners like most cars use. If you are changing the color of the exterior, paint the interior before you start the upholstery projects. Most trucks look best with the exterior and interior the same color. Some people paint the

When you have the entire interior stripped for a color change, check the condition of any areas that were previously undercoated. The undercoating inside this door needed some touchup, so a spray can of undercoating was used. Watch the undercoating overspray.

Painting Interior Parts

Cars with center consoles need to have them removed in order to install new carpet. Be careful when trying to remove a console because the fasteners can be hidden or difficult to reach. This Camaro console was mounted on three metal braces that are attached to the transmission tunnel.

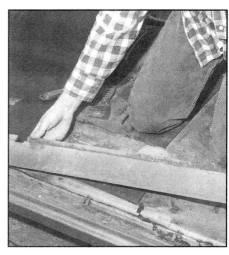

When you remove the old carpeting you might be surprised at the poor condition of the floorboards and related parts. On older Chevys the door sill plates are prone to rust. The screw heads can be rusted or stripped. If so, you will need to drill out the screw heads.

Some parts may seem very obvious as to their location when you are taking the car apart. Don't trust that your memory will be so accurate several weeks (or months) later when you reassemble the interior. Chalk can be used on the back side of parts to mark them. You can also make arrows to indicate the correct direction of the part.

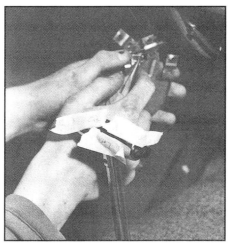

In order to paint a dashboard, all the items attached to it should be removed. That can mean a lot of wires and cables. Mark each item with a piece of folded over masking tape as it is disassembled.

are pretty self-explanatory as to where and how they are attached. But items that have several different types or sizes of fasteners may require a factory shop manual or a good aftermarket one.

Record Everything

Taking a few notes and making some simple diagrams as you remove each item is cheap insurance. Even better is taking some photos or using a video camera if you have access to one. Another tip is to remove parts one at a time and mark them with a piece of masking tape as they are removed. This technique works well with headliner bows, for example.

Even though it may seem obvious as to where the parts go as you remove them, many little parts and fasteners look very similar. I like to use the sealable clear plastic sandwich bags for storing parts, with a note in each bag explaining where the parts belong. The sandwich bags also prevent the tiny fasteners from getting lost. Put the collection of sandwich

interior parts black which will work well with most exterior colors.

Painting the interior of a truck can be quite time consuming. There is a lot of masking and sanding. A poorly prepped surface will shorten the life of your paint. I learned this the hard way when I did a quickie prep job on a bright orange truck that I changed to black. I just scuffed the orange interior sections and shot black over them. I skipped the primer step. It wasn't too long before scratches and chips revealed the orange color underneath. The cab looked like a bad Halloween decoration.

DISASSEMBLY

If you are going to paint the entire interior, you'll need to disassemble and remove as many components as you can. Many assemblies and pieces

Painting Interior Parts

Often there will be some type of insulation or mat under the carpet. Chances are great that this material is wasted and hard to remove. A putty knife will help get the old material off the floorboards.

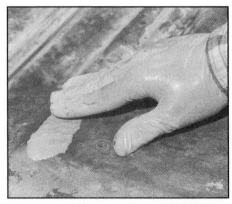

If you have minor rust pitting or pinhole rust, repairs can be made easily with POR-15 two part epoxy. The putty is mixed and pressed into the affected area. After it dries, it can be sanded.

There are rust preventative products like POR-15 that can be used to protect the floor from any future problems. Before the rust preventative paint can be applied, the metal needs to be chemically etched and rinsed.

bags in a box. You can also use various other household containers like jars with screw-on lids, margarine tubs, salted nut cans, and plastic cookie trays to organize parts.

It is possible to remove the seats and recover them before proceeding. Although you can restore separate items of the interior one at a time, in this chapter we are going to talk about total disassembly.

Totally disassembling the interior first decreases the chance of damaging any new parts while you are painting or cleaning. If you can keep the new parts separate from the work area until they are needed, you greatly decrease the chances for damage.

A large work area is a real plus. This is especially true if you take apart the steering column for repainting. If you have all-around good access to the car, you don't have to bother using a floor jack to move it. The work is just easier in a roomy environment.

Take care when removing old fasteners. Stripped, frozen, rounded, or broken fasteners can take a lot of time to extract without harming the surrounding surfaces. Use the proper tools on stubborn fasteners and at the first sign of trouble, back off. Study the problem rather than just trying to muscle your way through.

When you remove old parts, save them until the job is through. You might need them for reference. An example is with an old headliner. You can see where any notches were cut to help the fabric fit in corners. Don't throw away any fasteners or clips, not even ones that are less than perfect. Many of these little parts are difficult to find, especially late at night or on a weekend. Items like door panel clips that have lost some of their holding power can often be repaired if you don't have any new ones.

A shop manual is very helpful if you decide to disassemble the dashboard or steering column. Components for these items are often secured in unconventional ways or with hidden fasteners. People often use brute force when they just missed a little clip or set screw. If you disassemble the steering column for painting, have the wheels straight before you start so that you don't reassemble the column and find that the steering wheel is off center.

If you want to repaint the metal parts of the dashboard or change the color, it is good to disassemble it as much as possible. You can tape off the gauges and knobs, but this approach often leaves little bits of the old color exposed or gets paint on the gauges and knobs. Taping works better for repainting a dash the same color than for changing colors.

CLEANING AND RUST PROBLEMS

Once the interior of your car has been completely removed, thoroughly clean the area. Use a shop vacuum to pick up the years of accumulated dirt, dust, and left-over French fries.

Painting Interior Parts

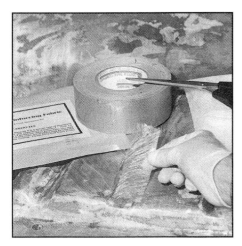

Actual holes (that aren't too big) can be patched with POR-15 steel reinforcing fabric. Duct tape should be applied to the bottom side of the hole and left there until the patch has cured. The fabric can be cut to size with common scissors.

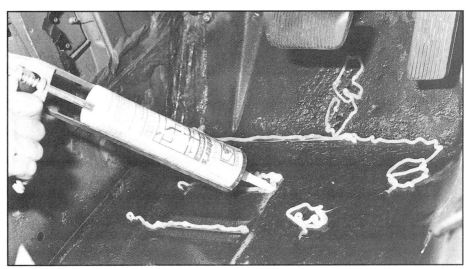

Whenever you remove old carpet, check the condition of the floorpan seams. If they are in poor shape, use a caulking gun to apply new seam sealer. You don't have to do a perfect job since the sealer will be hidden by the carpet. The goal is to keep moisture out of the interior.

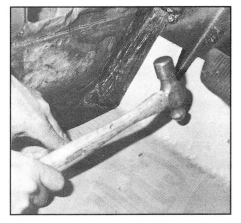

If your floorboards have large areas of rust out, they will need to be repaired. There are replacement floorpan sections available, but they require welding. If you aren't building a show car and just want to keep your feet off the ground, a patch can be made out of a piece of sheet metal. An old metal sign was used here. The bend was made with a ballpeen hammer.

The budget no-welding patch panel was secured with rivets. Before you drill any of the rivet holes, check underneath the car to be sure that there isn't anything in the way. Any patch panels should be sealed with seam sealer or clear silicone sealant. Lay down a nice thick bead to insure good coverage.

Various painted metal surfaces should be cleaned even if they are going to be repainted. General all-purpose cleaners will work, as will automotive wax and grease remover. Whatever product you use, take care not to leave lint or paper towel residue since it can affect the quality of any new paint.

If items like your headliner or dashpad are in good shape and you don't plan to replace them, now is a good time to give them a thorough cleaning.

When you remove the carpet, check for signs of water damage. You may have noticed this problem previously. If you have procrastinated about repairing the leak, you should take care of it now. It doesn't make sense to let a leakage problem ruin your new carpet. The most common sources of interior water damage are defective weatherstripping, a leaking heater core, and cowl vents that don't seal properly. Cowl vent problems can be a real pain, but they are much easier to fix when the interior is apart. One of the Mustangs used in some of the photos had cowl vent problems which had led to minor floorpan rust.

Serious rust-out problems mean that you either need to take your vehicle to a body shop or learn how to weld in patch panels. If you are lucky and only have minor surface rust, that can be easily fixed. A wire brush or a wire

Painting Interior Parts

Interior surfaces like Mustang doors shouldn't be heavily sanded. There is a fine pseudo grain that can be removed if you sand too vigorously. You just want to dull the original paint so a Scotchbrite pad will do the job.

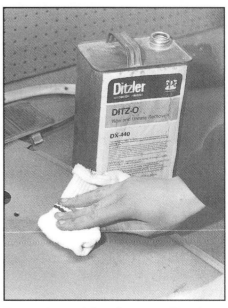

Exposed metal interior parts like these Mustang rear seat side panels should be thoroughly cleaned before you do anything else. A good quality wax and grease remover should be applied. By cleaning first, you avoid grinding any grease into the metal when you sand.

wheel attachment for an electric drill can be used to remove the rust that commonly forms in the corners and recessed areas of the floorpan. You need to grind the rust down to bare metal and protect the area with a rust inhibitor or metal prep solution. These products can be found at an auto body supply store. The whole area should be coated with a quality primer and then painted. Even though the main part of the floorpan will never be seen, it is a good idea to paint the outer edges the same color as the door jambs to give that factory original look.

When the gutted interior is clean and free of any rust or leakage problems, you can move on to painting all other metal surfaces.

PAINT BASICS

I highly recommend that you obtain any number of the auto body books available from HPBooks, such as the *Paint & Body Handbook* or *Automotive Paint Handbook*, to help give you the necessary background for painting the interior of your car. In this section, I will provide you with some basic tips and techniques that will help you get started.

Prep

The key to good interior painting is the same as good exterior painting—preparation. All surfaces that will be painted need to be thoroughly cleaned. The old paint needs to be sanded or scuffed for proper adhesion of the new paint. Unless you have a dented dashboard, you shouldn't have to deal with any sheet metal repair. If you do have a dent or any scratches deeper than 1/8 inch, they will need to be filled just like exterior dents.

Most shallow scratches can be block-sanded so that they blend into the surrounding area. If a scratch is still visible after some primer has been applied, use spot putty to fill the scratch. After the spot putty dries, block-sand the area again. The repaired scratch should be "featheredged," which means that the spot putty gradually tapers off into the surrounding metal. When you spray more primer over the repair, it should be undetectable. If you can still see

Since you are doing the job yourself, you can take the time to do extra work that you probably wouldn't want to pay for at a shop. The inside of these doors and the rest of the hidden metal surfaces were painted with POR-15 black rust preventative paint.

the problem area with primer, it will be visible under the new top coat of paint.

Painting Interior Parts

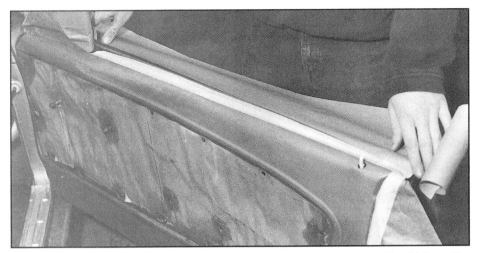

After scuffing the surface, clean it again with wax and grease remover. Use high quality automotive (not the weaker household variety) masking tape and masking paper to thoroughly mask off all areas not being painted. If you are painting a dashboard, use extra care when masking. You don't want any paint overspray to get on gauges or light bulbs.

The last step before applying any paint is to wipe down the surface with a fresh tack rag. The tack rag is sticky so it will pick up any remaining sanding residue or dust.

The area around the ignition, either in the dashboard or on the steering column, is commonly scratched. Painting over the scratches isn't really advisable. The area must be sanded smooth and brought up to the same level as the surrounding paint. Otherwise, the scratches will likely show through.

If you are painting the interior parts the same as they were before, you don't need to do as much prep as you would if changing the color. Check with a knowledgeable paint store or restoration supply firm to make sure that you get the right paint. Not all interior surfaces were painted glossy colors. Some had varying amounts of flattening agents added to the paint. It doesn't really matter on a daily transportation car whether you have the right percentage of flatness in your black paint, but if you can get the correct paint, go ahead and use it. The correct color paint sprays the same as an incorrect color.

When painting over the same color on metal that isn't rusted or otherwise flawed, you can clean the metal and lightly scuff the old paint with a Scotchbrite® pad. Clean the metal again after scuffing with a fresh tack rag. Using a tack rag is especially important when you are changing colors and when sanded coats of primer are involved. Painted interior surfaces aren't as critical as exterior areas, so the amount of care you invest depends on how particular you are about the final results. Even an outstanding interior paint job only involves a fraction of the work needed for exterior painting, so a little extra care will yield a superior job.

Masking

Removable parts won't require much, if any, masking. If the window frames have felt whiskers in them, mask them. The same goes for any rubber weatherstripping. Nothing says "repaint" more than overspray on areas that aren't supposed to be painted.

Most removable interior parts that need painting are relatively light, so I like to hang them from the shop rafters with wire. The parts are often unusual shapes so hanging them makes uniform access easier. A big piece of flat cardboard works well to contain the overspray around the part being painted.

When it comes to masking off the interior, you can't be too careful. Play it safe and leave a one-foot border of masking tape and paper. Many home painters use common newspaper, but I prefer the real masking paper. This paper is designed to repel paint and not soak through. It is also more pliable than newspaper. A one-foot-wide roll isn't very expensive and it will work great for interior masking. The paper is readily available at automotive paint stores.

Many people try to save a few dollars by using cheap masking tape. This is false economy. Use 3/4-inch masking tape that is designed for automotive paints. Go buy a fresh roll of tape right before you start the project. Old masking tape often gets brittle and loses its adhesive qualities. Good masking tape is essential if you need to tape off any rubber weatherstripping. It's almost impossible to get cheap tape to stick

Painting Interior Parts

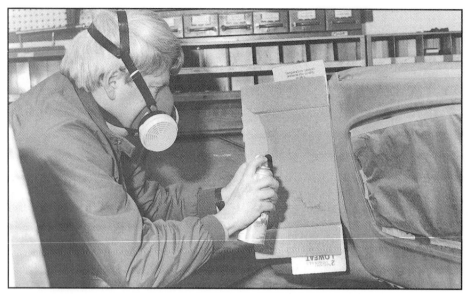

Even though you have masked all the surrounding areas, it doesn't hurt to be extra careful. Use a piece of cardboard to shield the areas you don't want painted. A paint respirator is always a good idea when painting.

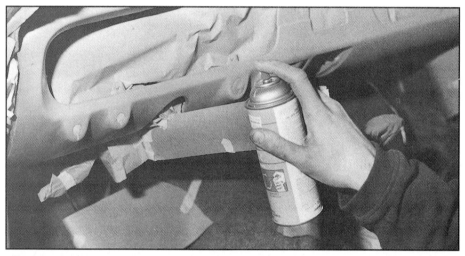

Changing the color of a dashboard requires extra care since there are so many odd corners. The steering column was removed for painting, but even so, the steering shaft was taped. The circular openings for switches were taped from the back side.

well to weatherstripping.

When you are masking the edge of an interior part, go slowly and use your fingernails to press the edge of the tape down. Tape that is hastily applied often lifts along the edges, which leads to unwanted paint bleeding. Use your fingertips to go over the taped edges before you apply any paint.

Removing Masking Tape—Just as applying masking tape correctly is important, so is the removal. Don't remove the tape until the paint is dry, but don't wait too long afterwards. Pull the tape off slowly to be sure that it isn't lifting any of the new paint. If the paint starts to lift, cut it away from the tape with a sharp, single-edge razor blade or an X-Acto knife. Don't let the masking tape get sun-baked or brittle. Tape that has been left on too long can leave a residue that is difficult to remove.

Dashboard Masking—It is always best to remove as many items as possible from the dashboard before painting. If you must leave gauges or switches in place, use extra care in masking. If the item to be masked is a shape that is hard to work with, mask it with more than one size of tape. Cover the border with either 1/8-inch or 1/4-inch tape first and then place the standard masking tape over the thinner tape. The thinner tape is much more flexible and easier to bend around corners. You can also get thin rolls of blue "fine line" tape which is even easier than thin rolls of regular masking tape to use in tight spots.

Applying Paint

Once the surfaces have been cleaned and prepped, you have several choices for applying paint. Spray cans are the easiest for inexperienced painters. Bulk paint is easier to get an exact color match, but it requires the use of spray guns. A spray gun means that you need a source of compressed air.

Spray Cans—If you are using spray cans, buy the best quality possible. Not only will you get better paint, but you usually get a better nozzle, which reduces clogs and sputtering. Spray cans should be at room temperature for best results (consult the can directions for specifics on your brand). I soak spray cans in a bucket of hot water for a few minutes prior to using and shake the cans vigorously. These tricks help get the best possible atomization.

Several light passes are better than one heavy pass with a spray can. This technique applies to spray guns as well. You don't want any runs. Parts that are painted off the car are easiest to paint when placed in a well-lighted

PAINTING INTERIOR PARTS

The odd shapes of many metal interior panels can make them difficult to paint. These Mustang parts were suspended on wire hooks from the garage door track. That way it was easy to evenly cover all the surfaces.

The professionals use mixed paint in a regular spray gun, but since most amateurs prefer rattle cans, here is a tip to insure better results. Place the spray can in a bucket of warm water for several minutes before using it. Shake the can thoroughly. The warm water helps the paint atomize better and go on smoother.

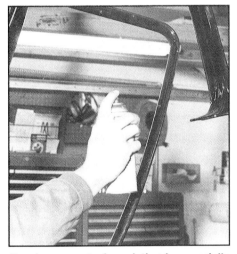

Hanging up parts for painting is especially useful when painting metal window frames. If you paint them on a flat surface, you can miss places where the metal is curved at the edges.

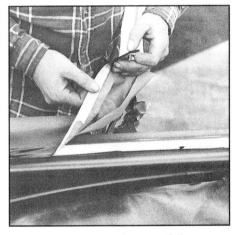

After the paint has dried, carefully remove the masking tape and paper. Notice how the tape is being pulled back over itself. The goal is to avoid pulling up any paint with the tape.

area. Spray can paint jobs are frequently marred by unevenly applied paint. Use a trouble light if necessary to check that you have adequate coverage.

Spray Guns—It is a lot more trouble to mix bulk paint, but this is what professionals use when they repaint interiors. I like to use a small touch-up spray gun for interior work. These small spray guns are easy to handle and they produce a smaller, easier to manage spray pattern than a full size spray gun. A beginner can get good results with very little practice using a touch-up gun.

I have also used airbrushes for interior painting, but they have very small spray patterns. A pattern that is too small can make the area look streaked. However, airbrushes are perfect for fixing scratches. Their fine control makes it easy to feather-in paint. Some spray gun and airbrush companies make a gun that is a cross between an airbrush and a touch-up gun. The nozzle style is like a touch-up gun, but it is much smaller than a standard touch-up gun and the paint cup is also much smaller. These work very well for interior work if you have access to one.

Airbrushes and touch-up guns don't need a lot of air like a full size production spray gun. This means that they can be run on most home air compressors. The smaller touch-up guns and airbrushes can be operated on little diaphragm style compressors.

Regardless of what type of paint you use and how you apply it, the keys to a good job are cleanliness, good surface prep, and careful application of the color coats. Be sure all freshly painted parts are dry before you reinstall them and start working in the surrounding areas. ■

Replacing Headliners 4

Headliner replacement falls into the old good news/bad news category. The good news is that there are excellent headliner kits available, they are very affordable, and few people spend their time staring at the ceiling of your vehicle. The bad news is that headliners are one of the more difficult items to install if you don't have much experience. This isn't to say that you should shy away from installing a new headliner; you just need to exercise a little extra care when doing so.

The problem with headliners is that to a achieve a professional quality installation, the headliner must be stretched as taut as the proverbial drum. Getting the basic headliner in place isn't difficult at all. The tricky part is making it tight and wrinkle-free, especially around the edges. The degree of difficulty has a little to do with the style of your headliner. The more curves and hard-to-reach corners your car has, the more difficult it will be to get a perfect fit. Of course, many newer vehicles have molded, snap-in headliners. Things don't get much easier than this. You remove any dome lights and/or clothing hooks; drop out the old headliner; snap-in the new replacement unit (apply a little glue if required); and reinstall any lights or trim pieces. Installing a traditional, stretched-fabric-type headliner requires a lot more work.

The edges of traditional headliners can be difficult to secure properly. Some cars require the removal of the windshield and/or rear window in order to wrap the edges of the headliner around the perimeter of the window opening. A quick look at your headliner should give you an idea of the potential difficulty. A detailed factory shop manual (which we highly recommend for any restoration project) will show you where any hidden fasteners are located and tell you how your headliner is secured to the vehicle's roof.

A general rule about headliner replacement difficulty is: the older the vehicle, the more difficult the headliner. By older, we mean fifties and before. Some old cars used screws, tacks, tack strips, glue, and metal grip teeth and any combination of these fastening devices. A shop

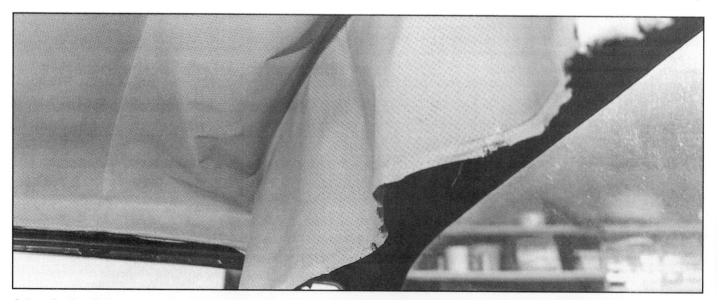

Automotive headliners are usually made of fairly lightweight material, so it is not uncommon for them to get damaged over the years. Light colored headliners can get quite dirty, too. Headliner kits are very reasonably priced. Some vehicles (like early Mustangs) require the removal of the windshield and rear window in order to install a new headliner. Removing and replacing the glass is a little difficult, but the headliner is easier to work on with the glass out of the way. All the chrome trim must be removed first. Carefully lift out the glass so as not to crack it.

Replacing Headliners

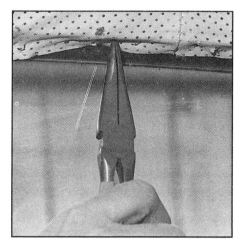

When all the molding pieces are removed you can see how the headliner was secured at the edges. Small staples are used around some back windows. Use a needle nose pliers to remove the staples from the tack strip. Many headliners are glued around the edges. In these cases, the headliner will usually rip as it is removed.

The original dome light lens and trim ring were missing on this '56 Chevy, so we ordered new ones from the C.A.R.S. catalog. Fortunately, the dome light base was still there since they aren't currently being reproduced. To remove the base, take a pliers and gently straighten the three tabs so that the base will slide over the tabs.

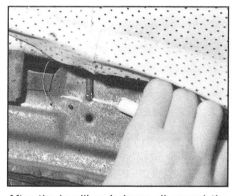

After the headliner is loose all around the edges, push back the headliner material to expose the ends of the headliner bows. Note that in most cases there are more than one mounting hole. Use a piece of chalk to mark which hole the bow was in.

Another option is to use pieces of masking tape to cover the unused holes. If you reinstall the headliner bows in the wrong holes, the headliner won't fit correctly.

manual will give you a clear picture of the intricacies of particular vehicles. Newer vehicles mostly used headliner bows and glue until the advent of the super easy snap-in headliners.

HEADLINER REMOVAL

For the most part, headliner removal is pretty self-explanatory after you remove items like window moldings. Although only the bows will be reused with your new replacement headliner, care should be taken to leave the original intact as much as possible. Occasionally there isn't a reproduction headliner available for your vehicle, so you'll need the original intact for a pattern. Even if there are reproduction headliners available, you may want the option of having one made from custom materials, so keep it in one piece. And finally, you'll want to keep the original to use as a reference for things like where to make relief cuts in the corners, or to see how far back the bow listings were trimmed.

Your chances of keeping the original intact are greatly increased if you use a factory shop manual. The shop manual will show you where, how, and in what order the headliner fasteners should be removed.

Remove Trim

Any window moldings or trim pieces should be removed while paying careful attention as to how they fit in the car. Most pieces are unique to a certain location, but some look similar and sometimes you need to deal with overlapping pieces. When there is any doubt, make a sketch or take a photograph. If the trim screws are all the same size and design, you can put them in a single container. If the fasteners are different, either put them in separate plastic sandwich bags or tape them to the back side of the corresponding molding.

Windlace—Depending on the type of car or truck you have, there is often some type of windlace around the door openings. If there isn't any windlace, then the trim panels serve the same purpose. The windlace helps keep out drafts and secure the edges of the headliner. The windlace needs to be removed. Sometimes it can be reused, but often it is faded and too stiff or too loose. Stiff windlace is difficult to work with. When windlace loses its clamping ability, it won't stay in place or do much to help secure the edges of the headliner.

Miscellaneous Items—Items that attach to the headliner like the rearview mirror, sun visors, dome lights, and garment hooks need to be

Replacing Headliners

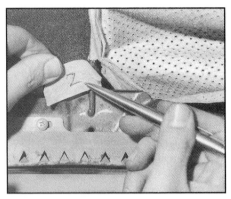

Although headliner bows are often the same or similar, some bows can be different. To be safe, place a little piece of masking tape around one end of the bow and number it according to its location from the front of the car. Remove the bows starting at the front of the car. The bows are held in place by tension. Either pull the bows out of their holes or push the whole bow backwards to relieve some of the tension.

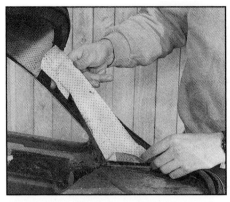

Many older vehicles, like classic Chevys, have extensions of the headliner that are separate covered pieces of panel board. The rear window was out of this car because it was broken, not because it needed to be removed for the headliner installation. The missing window did make access to the rear of the headliner much easier.

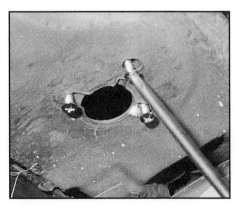

Many installers like to put the sun visor screws back in their holes after the headliner has been removed. This way they don't have to poke around to find the holes. They simply make small slices to expose the screws.

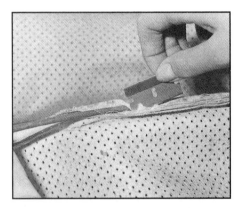

When you have the headliner out of the car and on the workbench, remove the bows. It is highly likely that the bows will be rusted to the listing. If that is the case, use a razor blade to cut the bow out of the listing.

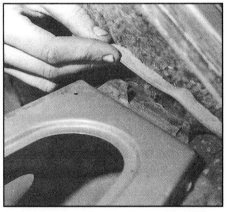

Many cars use a strip of panel board to secure the rear lower corners of the headliner. Don't lose these strips when you remove the old headliner. If they are water damaged, use the old ones as a pattern for new strips.

removed. The dome light screws are commonly located underneath the lens cover. The lens covers snap in place so a little gentle pressure around the edge of the lens should release it. If you haven't disconnected the battery (a good idea since you will have the doors open while working on the interior), remember that the dome light bulb can get quite hot and burn your fingers. After the dome light housing is removed, tape off the ends of any exposed wires.

There are two schools of thought as to the handling of the fasteners for these items. One method is to remove the screws and put them in a marked container. Then after the new headliner is installed, use your fingers to find the attachment points and carefully poke holes with an awl.

The alternate method involves replacing the screws as soon as the sun visor or hook is removed. The existing hole in the headliner is probably large enough that the screw will pass through the headliner. If not, use an X-acto knife to make the opening a little bigger. Then the retaining screws stay in their mounting locations while the new headliner is installed. When it comes time to mount the accessories, a small slit is made in the new headliner to reveal the screw. The sun visor (or whatever) can be installed without searching around for the screw location.

Both methods have their advocates. The important thing is to avoid making unwanted holes in the headliner or holes that are too big.

A trick to help you find the screw holes (when you choose to remove the screws) is to mark the approximate location on the outside of the car with a piece of masking tape. For example, the location of a garment hook can be marked with masking tape placed on the rain gutter.

Rear Package Tray—The rear package tray will often need to be removed. Aside from improving package tray access, the upper corners of the rear seat often hide the lower attachment points of the headliner. These points are sometimes literal

Replacing Headliners

Use steel wool, sand paper, or a Scotchbrite pad to remove the rust from the bows. The bows need to be perfectly smooth and clean in order to smoothly slide into the listings of the new headliner. It is not necessary (or recommended) to paint the bows. Carefully insert the headliner bows in the listings of the new headliner. Work slowly so that you don't snag the listing, which can easily be ripped. Work on a clean surface to keep the headliner free of smudges.

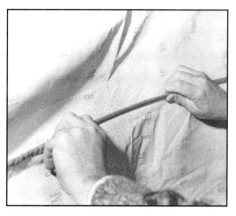

As you install each bow, check to see that the listing doesn't bunch up. Use your hands to smooth out the listing from the center outwards. This is why it is important to have smooth, clean bows.

(serrated metal "teeth" that the material is wrapped around) or tack strips or fold-over metal tabs.

Removing Windshields—If your vehicle requires that the windshield and/or rear window come out for headliner replacement, use extraordinary care. This is a two person job with one person pushing from the inside and the other outside to handle and catch the glass. Old glass and very cold glass are especially sensitive to cracking. If you have any doubts about your ability to remove the glass intact, do the rear glass first. It is usually less expensive than the windshield if you damage it. When in doubt about the glass, have a professional handle the job.

The headliner is usually glued to the roof just behind the windshield. You will need to peel the headliner away from the roof. The headliner is often glued around the edges and rear of the roof also. When these edges are pulled loose, the headliner should hang free around its perimeter.

Weatherstripping—Any time that you remove glass, you will probably need to buy new rubber weatherstripping. Even though the old weatherstripping may be serviceable, it makes more sense to install new stripping with all of the glass out.

Headliner Bows—Now it's time to remove the headliner bows, if your car is so equipped. I recommend that you start at the front and work toward the rear. The bows are usually secured to the inner framework of the roof by screws or holes (or sockets or indentations) that hold them taut. The snap-in bows are kind of like tent poles. You need to release the tension to remove the bow. Sometimes you can push up at the curved section of the bow with the palm of your hand while you pull out the end of the bow with your other hand. Some bows can be released by pushing the bow toward the rear of the car.

Sometimes there is more than one possible hole for the bow ends. It is important that you mark which hole was used as you remove the bow so you will remember where it goes when you install the new headliner. The slot can be marked with a marking pen, or take a little strip of masking tape and cover the unused holes.

It is also very important to mark each bow as it is removed from the headliner. You can number the bows from front to back and put the number on a piece of masking tape. Another technique is to make a dot with a marking pen; one dot for the first bow, two dots for the second bow, etc. The bows may appear similar, but they are different and must be reinserted in their proper location if you want the new headliner to fit properly.

The last bow is often held in position from the rear as well as the side bow slots. This is usually done by one or two "wires" that hook over the center of the bow. The other end of the wires are secured to the back window opening. Don't forget these little wires when it comes time to install the new headliner.

After the marked bows are out of the vehicle, check to see that they are clean and rust-free. Rusty bows can stain new headliners, especially white or light colored ones. Avoid any future problems by cleaning the bows with some steel wool or a Scotchbrite® pad. If the bows were really rusted, you may want to paint them with a rust preventative type paint. Be sure the paint is completely dry before putting the bows in the new headliner.

With the headliner and bows out of the car, check the condition of the insulation material (if any was used on your vehicle). The insulation can sag with time. If the insulation is still in good shape, it can be re-glued with some contact cement. If it is badly worn or rust stained, replace it with new insulation. New insulation doesn't need to go all the way to the edge of the roof.

Replacing Headliners

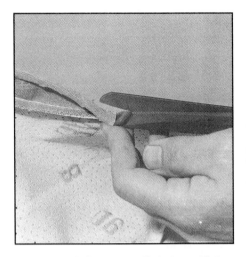

Often, the listings are a little long. If they cover up the ends of the bows, cut away the excess with a scissors. Otherwise the headliner will bunch up where the bows are inserted into the roof. Just cut a little listing at a time. You can always trim a little more after the bows are installed.

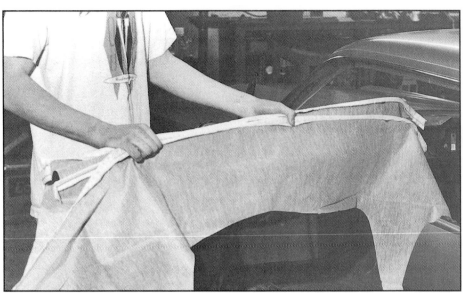

Fold up the headliner and bows like an accordion to put it in the car. Remember to be sure that your hands are clean. Start the headliner installation at the rear window, unless the shop manual for your car says to start at the front. The important thing is to pull the headliner from one end of the roof to the other.

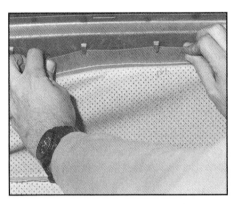

The rear of this C.A.R.S. '56 Chevy headliner has a cardboard strip just like the original headliner. This strip is secured by tangs along the rear of the roof, just above the rear window. Position the mounting strip as tightly as possible with the tangs. Be sure that the strip is centered.

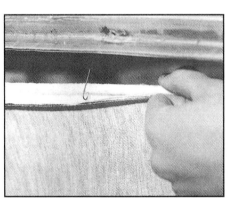

Some cars, like this Mustang, have small retaining wires for the rear headliner bow. The wires are attached to the inside of the roof. Hook them around the bow to anchor it.

HEADLINER INSTALLATION

Those experienced in upholstery make the installation of a headliner look pretty easy, often installing it by themselves in a short period of time. But beginners will most likely need an extra set of hands. The most important factor for installing a headliner is to get it stretched as tight as possible, which is obviously much easier if someone helps.

Lay out the headliner well in advance of installing it. Keep it in a warm room, or if the weather is good, place it out in the sun to warm up. Heat will help the packing wrinkles disappear.

For installation, I recommend that you work from the rear to the front, the reverse of the removal process.

Install Bows

Before you install the headliner bows, check to see whether the back side of the material is marked with a center line. It helps to put a small notch at the center line on both ends. This will help you center the material before securing it.

When you put the bows inside the listings sewn to the back side of the headliner, you will probably have more listing than bow. The excess will need to be trimmed if you want to avoid bunching at the edges of the headliner. Be careful not to trim too much listing or you could have the opposite problem of sagging near the edges. Trial-fit the headliner to be sure that it is centered and to determine how much listing to remove. You can also carefully trim the listing after the bows have been installed, but before any glue has been applied. Ideally, you would like only 1/2 inch to one inch of the metal bow exposed. You want the seam to be wrinkle-free, but to wrap nicely at the

Replacing Headliners

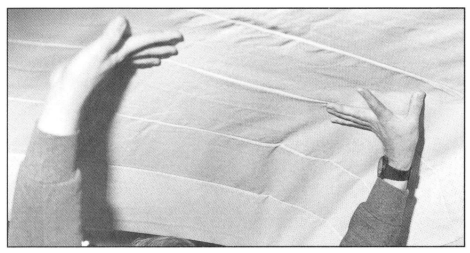

When all the bows are in place, use the palms of your hands to gently work the excess headliner material to the sides of the roof. If the headliner bunches up at the end of the bow, trim a little more off the listing.

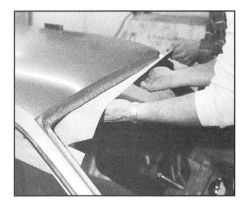

Contact cement is often used at each end of the roof. Spray the inner roof attachment area first. Then lightly touch the headliner to the brace. That will show you where the trim adhesive needs to be sprayed on the inside of the headliner. After the contact cement dries, pull the headliner as taut as possible before touching it to the brace. Pull evenly so that the headliner will be properly positioned. Use a spring clamp to help secure it if the windshield is out.

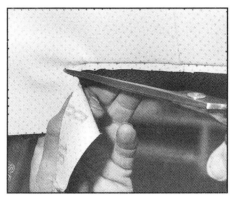

If you properly centered the headliner, you should have some excess material. Carefully trim the excess with sharp scissors. Be careful not to cut the headliner itself. Be very certain that you don't need the material before you cut it. You can always trim a little more, but you can't glue back any cut material.

edges above the window molding. This area is also known as the crown of the bows. You can take a look at the original headliner to see how the factory did it. This is another good reason to save the original headliner until you are completely finished with the job.

Put the rear bow into the correct slots. More care needs to be used installing the new bows than removing the old ones because it doesn't matter if you tear the old headliner. If there are hooks that pull the rear bow toward the rear window, remember to install them. Work your way forward, installing the rest of the bows. Check to see that the headliner is centered and trim the excess bow listings (if you didn't do it previously) as needed.

Spring clamps can be used to help position the headliner. They also work like an extra set of hands when it comes time to stretch and finish-fit the headliner. Don't be too anxious to trim the excess material. You want to wait as long as possible to do much trimming. If it is obvious that you have several extra inches, you may trim a little, but save the final cutting until the headliner has been stretched and glued. Remember the carpenter's axiom of measuring twice and cutting once. Trimming excess is easy; trimming too much means buying another headliner.

Securing Edges

When you have all the bows in their correct slots, you will have a very sloppy looking headliner. That is fine since the real fitting comes next. Double check that the headliner is positioned like you want it. Now you need to use glue to secure the perimeter.

Automotive trim adhesive, like 3M Super Trim Adhesive (#08090) or 3M General Trim Adhesive (#08080) is used to secure the headliner to the front, rear, and often the sides of the roof. This type of adhesive needs to be applied to both surfaces and then allowed to dry. It should not stick to your finger when tested. When the two surfaces are joined, the bond is very strong.

Rear—To determine where to apply the adhesive to the back side of the headliner, spray the glue on the edge of the rear window opening. Some vehicles have a tack strip or gluing strip at either end of the roof. Pull the headliner taut over the attachment area. Press the material to the glue strip briefly, then pull it back. You will see a mark where the glue made contact. Spray more adhesive on the back side of the headliner where this mark is. Then let the two

Replacing Headliners

On vehicles with the glass removed, fold the headliner around the opening and glue it in place. After the glue has set, a straight edged razor blade can be used to trim off the excess material. The rear corners are the most difficult part of a headliner installation. A spring clamp was used to hold this headliner in place while working on it.

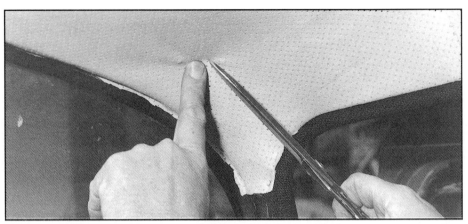

If the sun visor screws were placed in their holes before the headliner was installed, use your fingers to locate the screws. Then cut a small opening with a scissors or a razor blade. This is easier than trying to find the holes under the headliner.

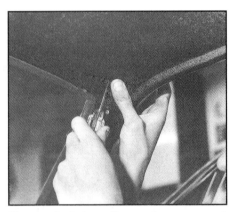

If you prefer to leave the sun visor screws out while the headliner is installed, you will need to find the mounting holes with the tip of an awl. Try to feel the holes first with your fingers so that unnecessary holes aren't made.

surfaces dry before mating them.

When you mate the headliner to the glue strip, start in the middle and work your way out to the edges. Pull the headliner tight to the back window and use the palm of your hand to work the material to the outer edges. This is a situation where a helper can pull the headliner around the window opening while you work out any creases or wrinkles from the inside. When the headliner is taut, clamp the outer edges until the adhesive sets.

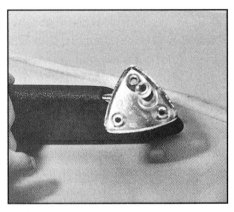

Always save the old headliner as a reference source until the new headliner is completely in place. In this case, the location of the sun visor screws and pivot post were compared to give a better idea of where to cut the new holes.

A trick you can use to help position and secure the headliner edges during installation is to cut up part of the old windlace (it must be the snap-on style windlace) into short sections. Use these three- or four-inch sections of windlace like mini clamps.

Front—The same glue trick needs to be employed at the front of the vehicle. Pull the headliner as tight as possible before you apply the glue and again when the two surfaces are mated. The tighter you get the headliner now, the better the finished job will be. You can remove some wrinkles with steam or dry heat, but reserve that lifesaver for corners and small problem areas. The main, flat area of the headliner should be as tight as possible without any rescue operations.

Sides—After the headliner is taut from back to front, it is time to tighten and secure the sides. Again, work from back to front and side to side in each area. The palm of your hand works well to smooth out wrinkles while you tug on the edges with your other hand. While all this tugging and pulling is going on, check to see that the headliner seams remain straight.

Notching Corners

As you are working from side to side at the rear of the car, you will encounter one of the most difficult parts of the job. The compound curves at the back of many roofs mean that small relief cuts are necessary to make the material lay as flat as possible. Look at the original headliner to see how the notches were cut. Work very carefully in this area. You want the minimum cut that will relieve the tension just enough to make the corner smooth. Too big a notch will take the shape right out of the corner and possibly extend past

Replacing Headliners

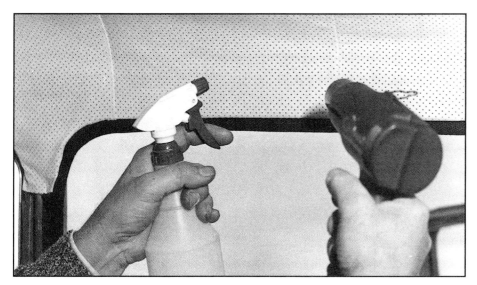

A spray bottle of water can be used to alternately heat and cool the wrinkled area. Go extremely slow with this process. This is a home version of using steam to reduce wrinkles. If you don't feel comfortable about removing wrinkles, consider taking the car to a professional who can use a steam gun to remove wrinkles.

the area covered by the window moldings. Besides making the corners flow smoothly, you don't want a bunch of folded headliner material at each corner. This part of the job is akin to a perfectly wrapped present.

It is much easier to notch and secure the corners when the glass has been removed. If you replace your headliner with the glass intact, you will need to be more careful with the corners. They can still be relieved and glued in place, but remember that the notches must be hidden by the rubber window molding. A dull putty knife or a special upholstery tucking tool will help get the edges of the headliner up inside the weatherstripping.

If the sides of your headliner require gluing, use the same two-part technique as with the front and rear window areas. The sides of the headliner are secured and finished with new windlace molding. This material comes in rolls and you start at a bottom edge and push and roll it into position. The inner wire reinforcing can be tight, but that is how you want it. After the snap-on windlace is in position, use a soft-faced plastic or leather mallet to make sure it is all the way on. A regular hammer can mar the windlace. You can also use a block of soft wood to distribute the force of the hammer blows.

If you have a little extra material hanging over the edges, trim it before installing the windlace. Leave only as much material as will be covered by the windlace and trim moldings. The more material that is wrapped around corners and edges, the better it will hold.

Removing Minor Wrinkles

No matter how hard you pull and tuck, there might still be a couple minor wrinkles. Professionals remove wrinkles with a steam wand. You can spray a light mist of water from an adjustable spray bottle and dry it with a heat gun or hair dryer. The combination of water and heat will shrink the area, but avoid using too much water or heat. It is better to go slowly and work over an area several times than to go too quickly and cause permanent damage.

You can also work out wrinkles with heat and no water. Steam is most often used on cloth headliners, so heat alone is fine for vinyl headliners. A hair dryer works well for beginners. It is cooler (and slower) than a heat gun, but less apt to cause damage. Whenever you use heat, always keep it moving. Don't focus closely on one spot. That is a sure way to damage the material.

If you are in doubt about using heat to remove wrinkles, practice on a scrap of material that was left over from trimming. Purposely wrinkle the scrap and then use the hair dryer or heat gun to make it smooth again.

Reinstall Windshields

When you are satisfied with the fit of the headliner, you can install the front and rear windshields if they were removed. And, as already recommended, use new rubber weatherstripping and sealant caulk.

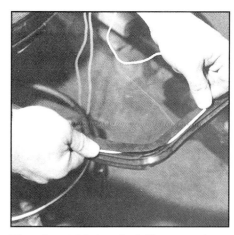

There is a trick to reinstalling auto glass. Place a new gasket around the glass. Then place a cord or vinyl covered clothes line in the recessed part of the gasket. Position the cord so the two ends meet in the bottom center of the gasket.

Replacing Headliners

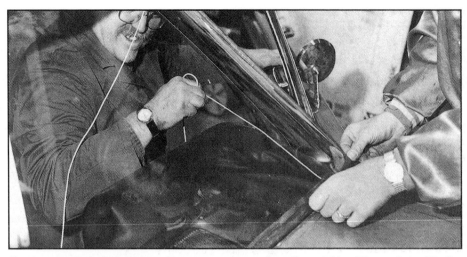

Use two people to install glass. Have one person push on the outside of the gasket while the inside person pulls out the cord. When the glass is properly seated, the stainless steel trim pieces can be snapped back in place.

Some cars use a U-shaped snap-on trim strip around the window openings to help secure the headliner and hide the edges. This trim has stiff wire inside it so you may need to use a rubber mallet to get the trim all the way on.

Installing the painted window garnish moldings is one of the final steps on cars equipped with these moldings. New, correct mounting screws are available from reproduction companies such as C.A.R.S. New window whiskers were installed on the window frames after they were painted.

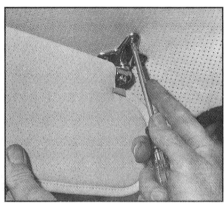

I avoided a lot of aggravation by ordering the pre-sewn Chevy sun visors from C.A.R.S. I also got new mounting brackets, a reproduction rear view mirror, and a dome light bezel and lens.

You can make the installation easier if you wrap a continuous piece of sash cord around the glass. Insert the cord in the center channel of the rubber. Make the two cord ends meet and criss-cross at the bottom center of the glass. Position the glass with a helper holding the glass on the outside while you slowly pull the cord out of the rubber. Pulling on the cord will make the rubber slide over the metal lip of the window opening.

Install Miscellaneous Items

The last step is to reinstall all the accessories like the rearview mirror, sun visors, dome light, and garment hooks. If you left the screws underneath the headliner, find them and carefully cut a tiny hole to expose them. If you removed the screws, find the mounting locations with your finger and puncture the headliner with an awl. Install the accessories and the metal moldings, and you're finished. ■

Here is the finished reproduction headliner in our '56 Chevy Delray. It looks just like new and is an incredible improvement over the old headliner.

CARPET 5

Replacing the carpet in your car or truck is one of the easiest upholstery tasks. There are many excellent companies that make replacement carpet sets for most vehicles. The basic routine is to remove the old carpet, clean the floorpan, and install the new carpet.

Most of the carpet available today is very close to the original materials and colors. You may have a little trouble if your car is an unusual color. Some reproduction carpet is made out of different materials, but the look is usually the same as the original. Authenticity is really only a problem if you are restoring a car to the factory specifications and are worried about judging points. If you are this particular, you should spend some time researching the various reproduction carpets to find the most authentic one. Areas to watch for include the height of the pile and the size of the heel pad.

Many carpet companies will send you a small sample of the type of carpet that they sell for your car. You can use this sample to see what type and quality of carpet you can expect. You can also compare the sample to your original carpet. Quality can vary between different companies, so don't shop price alone. If you have a local source for carpet, you can examine the carpets before you buy. The variety of factory-made carpet available for your vehicle depends largely on the popularity of your particular model.

If your carpet is ripped, stained, and moldy like this thrashed '56 Chevy carpet, it is time for a replacement. No amount of cleaning can save a ratty carpet like this.

Carpet

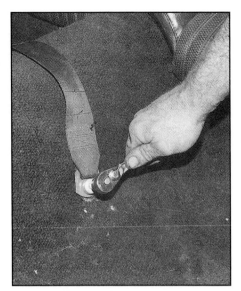

Most vehicles need to have the seats removed in order to remove and replace the carpeting. Cars like early Mustangs have the seat retaining nuts in the floor pan. Rubber plugs hide the fasteners. Another common type of seat fastener is accessed from the top side under the corners of the seat. Seat belts need to be removed in order to install new carpeting. Many seat belts are secured with Torx fasteners. They are similar to an Allen bolt, but with a "star" shape. You will need a proper size Torx socket to remove the bolts.

Anything that is mounted to the floor such as a center console or a shifter needs to be removed. Usually the top of the shifter has to be removed in order to slide the base piece up and over the stick. The handles are often secured with a small, hard to see Allen head fastener.

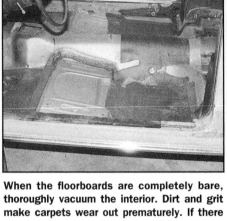

When the floorboards are completely bare, thoroughly vacuum the interior. Dirt and grit make carpets wear out prematurely. If there are any floorpan problems, repair them now. Refer to the photos in Chapter 3 for information on floor repairs.

TYPES OF CARPET

Your two basic carpet choices are *loop pile* and *cut pile*. The pile is also known as the *nap* of the carpet. The pile is what gives the carpet its texture or feel. Loop pile is when the threads are woven in loops and bonded to the backing material. When the ends of the loops are shaved, you get cut pile carpet. In either style, you may find high or low pile. The usual range is about 3/8-5/8 of an inch.

Automotive carpet is different from household carpet. You might get away with some household carpet scraps in your trunk, but don't use it in the interior.

The vast majority of reproduction carpet is *molded*. That means that it comes pre-fitted to all the high and low spots of your floor. Molded carpets virtually drop into place.

The other style of carpet is known as *cut and sewn*. Older cars used cut and sewn carpet. A sign of this style carpet is sewn edging. You will notice seams around the transmission tunnel and the firewall. If your vehicle requires custom carpet, an upholstery shop will have to make a cut and sewn carpet. It is possible for you to make your own patterns and cut out the carpet sections. You can then have the pieces sewn on a commercial sewing machine. Before you undertake such a task, check to see if a local shop is interested in doing just the sewing. They won't be able to guarantee the fit since they didn't make the patterns. Some mail-order carpet companies offer cut and sewn carpet in addition to molded carpet.

Older trucks and street rods are prime candidates for cut and sewn carpet. Owners of modified vehicles often desire premium carpet like that found in expensive German luxury cars. The only way to get this type of carpet (especially in custom colors) is to have it sewn. Even if you don't have the ability to sew your own carpet, you can do the floor prep work and apply any insulation material.

If you have a choice of carpet styles, consider how you use your vehicle and how much you like cleaning it. The deeper, plusher carpets tend to get matted with heavy use. They also trap more dirt and sand because of the longer fibers. Trapped dirt can increase the wear factor so you will need to clean it often with a powerful shop vacuum.

Quality

The density of the pile is a reflection of the carpet quality. Deeper, denser carpet is considered plush pile. Carpet samples can help you determine the grade of carpet that you will get.

If your goal is to refurbish your

Carpet

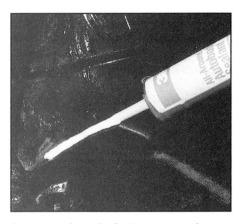

Any areas where the factory seam sealer was removed or otherwise disturbed should be repaired with fresh seam sealant. Your local auto body supply store can provide a variety of sealants in cans or tubes. We used a caulking gun to apply a uniform bead of 3M All Around Autobody Sealant #08500. The sealants that come in a can are applied with a small brush.

If you are in a hurry, you can just install the new carpet as it comes out of the box. If you want to make your interior quieter and better insulated, consider installing some heavy duty insulation. This Roadkill Carpet Pad RKC from Stinger Electronics is designed to make an interior a better environment for a high quality sound system.

vehicle, you may even decide to install a superior grade carpet or to add carpet in place of the factory rubber mats installed as original equipment on some cars. Of course, if you're building an all-out street rod, you'll want the finest grade carpet you can afford.

REMOVAL AND FLOOR PREP

If you have previously removed all interior components, you are in good shape. If not, you need to remove everything that is on top of the carpet, including seats, seat belts, door sill plates, and center consoles. Many seat belts require the use of a Torx socket. Center consoles are sometimes comprised of several interlocking sections. Often the consoles are made of fragile plastic, so don't force anything. A factory shop manual should provide an exploded view illustration to make disassembly easier.

Cleaning

When you remove the old carpet, check for signs of water damage. If you have unwanted moisture, you need to stop it before installing the new carpet. Soggy carpets under the dashboard on the passenger's side often can be traced to a defective heater core. Even small leaks should be repaired (you usually will need a complete new heater core) before they become a major problem. Some vehicles, such as early Mustangs, are known for leaky cowl vents. These problems are with older vehicles, and while you have the interior completely out of the car, now is the time to fix them.

This is also the time to replace or repair rusty floorpans. Badly rusted floorpans can be fixed with welded patch panels that match the original flooring. If this is beyond your capabilities, you will need to take your car to a body shop. If function is more important than form, you may be able to make some stop-gap repairs by riveting in pieces of sheet metal.

If you are lucky, you will only encounter mild surface rust. This should be cleaned down to solid metal with a hand-held wire brush or one powered by an electric drill. A shop vacuum should be used to clean the area.

A thorough cleaning job is important even if you don't have any rust problems. Prepare the bare metal with wax and grease remover and metal prep solution. A rust preventative paint should be used. Any place that the factory seam sealer has deteriorated should be renewed with a product like 3M Sealmaster or a similar seam sealant. Checking all the seams is a good way to prevent future moisture problems. When seams are resealed, allow ample drying time as specified (usually 24 hours) in the product directions.

INSTALLATION

Padding

Before you install the actual carpet, you may want to install insulation or add more to the existing pad. There are various types of carpet padding available. Traditional padding is often known as *jute padding*. Some upholstery shops use bonded foam padding similar to the padding used in houses.

There are also many brands of multi-layer insulation now available. These new high tech insulation products are a combination of an aluminized layer (or layers) and closed-cell foam or "bubble packing" type material. This "sandwich" construction works very well for reducing both heat and noise in the interior. You can also get noise reduction mats that are like heavy pieces of rubberized material. These

CARPET

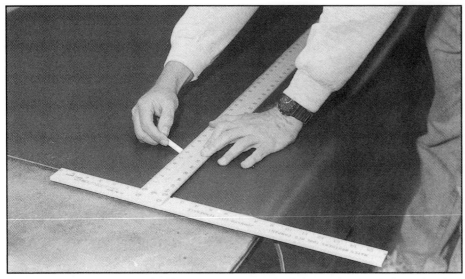

The Roadkill RKC pad comes in 52 x 48-inch sheets. The floorpan of a classic Chevy is wider than that so we had to do some calculations to determine how to make the fewest possible cuts. The best performance is obtained with the least number of gaps in the Roadkill RKC. We marked the pad with chalk and a T-square and cut it with utility scissors.

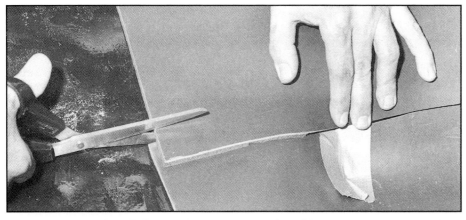

To make the Roadkill RKC fit nicely in the low areas of the floorpan, you will need to relieve the bulges. Push the pad into the foot well and then cut a straight line down the center of the bulge to the floor. Fold one edge over the other side of the cut to form a triangle that represents the bulge part of the pad. Remove that extra triangle of pad with scissors.

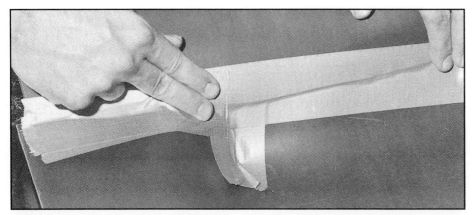

With the bulge cut out, the pad is flat and fits nicely into the floorpan. Use duct tape to join the two sides of the cut. When removing excess pad like this, it is always better to err on the conservative side. You can easily enlarge the cut.

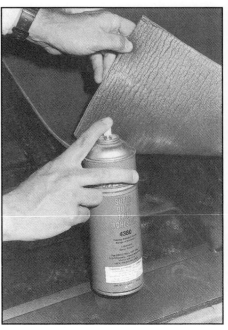

When you are certain that everything about the carpet pad is right, you can secure it to the floorpan with trim adhesive.

mats are sometimes called *underlayment pads*. You might find these pads under your factory carpet.

Cutting Padding—Regardless of which type of insulation you use (and you can use more than one type or layer for extra sound deadening) you will probably have to cut it to fit your vehicle. Some popular collectible cars have ready-made padding kits and many carpet kits come with padding already attached to the bottom of the carpet. However, the aluminized foam insulation comes in rolls that you must cut for a custom fit.

Because the insulation won't be visible, the fit isn't as critical as the carpet. A tip is to leave a one-inch gap between the various sections of the insulation. This leaves room for overlapped carpet sections to fit as flat as possible. This is particularly important if your carpet has bound edges. If the insulation is installed with proper gaps, the binding will rest flat in those gaps.

Carpet

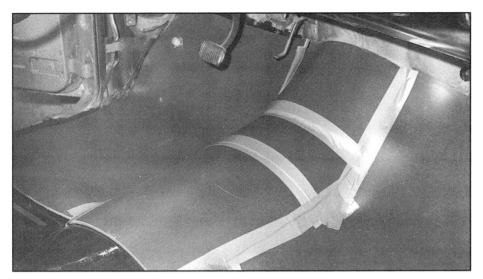

Here is the installed Roadkill carpet pad. One sheet was enough to cover up the edge of the front seat. Another sheet and a half are required to cover the whole floorpan. The most important areas to cover are those not covered by the seats. The seats help suppress noise.

There are a variety of insulation and padding materials available for automotive use. This car has giant wheel tubs which can transmit a lot of road noise. The area was first covered with foil-backed bubble style insulation to suppress both heat and noise. Then a layer of jute carpet insulation was added.

Quality reproduction carpets come with heal pads and shoe scuff pads already sewn in. The carpets may or may not have an opening for floor mounted dimmer switches. The center transmission hump area is not cut. You need to trim this area depending on the type of shifter you have.

When you use the aluminized insulation, it helps to make a pattern out of lightweight cardboard or heavy Kraft paper (or butcher paper). This insulation is a little costly, so patterns will help you avoid cutting mistakes and help reduce waste. The material can be marked with a felt tip pen and cut with a utility knife or heavy-duty shop scissors. This type of insulation should be glued to the floor (it can also be applied to door panels, kick panels, and firewalls). Many of the insulation kits come with adhesive that is applied with a caulking gun. You can use duct tape or aluminum foil tape to seal the seams between sections of insulation.

Many builders of street rods and older trucks use the aluminized insulation under a layer of traditional carpet padding for extra insulation. These older vehicles aren't as tightly constructed as modern cars and trucks, so the extra insulation can greatly improve noise isolation.

Securing Carpet & Padding

Carpet is commonly secured with either glue or nothing at all. Glued carpets look best because they won't slide or bunch up, and the fit can be tighter. But if you need access to anything underneath the carpet, you could be in trouble. Street rods in particular often have items like batteries or master cylinders that can only be accessed under or through the floorboards. Obviously, you can't be ripping up the carpet and insulation every time you need to check the brake fluid.

A solution would be to use strips of

Carpet

There are two schools of thought on carpet installation: those people who glue the carpet to the floor and those who don't. Gluing gives a better fit and the carpet won't slide around or bunch up. The problem with glue is that you can't remove the carpet for cleaning without damaging it. Trim adhesive should be applied to the floorpan and the bottom of the carpet. Follow the product directions and let the glue set up.

Carpet kits are usually made slightly wider than they need to be. This allows you to trim any excess, which can vary depending on whether or not a lot of insulation was installed. A utility knife or a single edge razor blade will do the cutting. Make sure that there is enough carpet to go under the edge of the door sill trim plate.

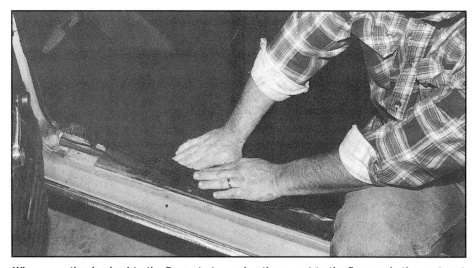

When carpeting is glued to the floor, start pressing the carpet to the floorpan in the center at the transmission hump. Carefully position the carpet toward the doors. If the fit looks correct, go back to the center and use your hand to get the carpet secured to the floor. Push the carpet toward the doors to avoid any wrinkles.

double-sided tape. If you pull up the carpet enough so that the tape loses its adhesion, you can simply install more tape. You can also use strips of Velcro tape. Custom upholstery shops often sew snaps into the corners of the carpet, and kits are available at upholstery shops. However, you'll have to have them sewn into the carpet.

Final Installation

While you are working on the insulation, lay out the new carpet kit in the sun (weather permitting) or in a warm room. The carpet is usually pretty stiff and the heat helps it relax from being packed so tightly in the shipping box. A pliable carpet is easier to fit and move around inside your vehicle.

Check the new carpet against the old carpet. Lay them on top of each other to be sure that you received the right carpet. Even though the box says it is what you ordered, mistakes can happen.

Cutting Holes—Trial-fit the new carpet to the car before you cut any necessary holes. Many people suggest using the old carpet as a guide for holes, but some shops warn against this. It is possible that holes in the old carpet may have stretched, or the floorpans may have sagged. This seems a little extreme, but you should check twice and cut once.

My solution for hole cutting is to use chalk to lightly mark the original locations with the old carpet as a guide. Then place the uncut new carpet in the car and use a sharp awl to double check by pushing up from the bottom side of the car through the hole. Once you are confident that the hole is in the proper location, use a razor blade to enlarge the hole with an "X" cut. Always make the smallest

CARPET

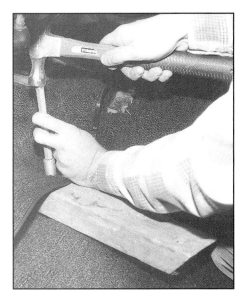

Some holes will need to be made for the seat fasteners and seat belt bolts. The idea is to make the smallest possible hole that the fastener will easily fit through. If the carpet is cut incorrectly, it can fray. Here, a gasket making punch is backed by a block of wood to make a hole.

Holes can be made with an electric drill, but drill bits are notorious for unraveling big strands of the carpet yarn. If you cut a small "X" with a utility knife and then place a small section of tubing over the cut, the drill will be less likely to cause any problems.

possible cuts and holes first; you can easily enlarge a hole, but you can't shrink one. Tight-fitting carpet around seats, shifters, and center consoles is a sign of a professional quality installation.

Once you have a bolt or screw hole cut, place the fastener in that hole. This trick will help keep the carpet in place as you locate and cut additional holes. When cutting larger holes for items like seat belt fasteners, be careful not to unravel the carpet. Some people use a piece of metal tubing that is slightly larger than a drill bit needed for the hole. They use the tubing as a guide for the drill bit. If you are in doubt about cutting big holes in the carpet, practice on your old carpet.

Trimming Excess—Most reproduction carpet kits come slightly oversized. Don't be too eager to trim the excess around the door openings. Even though the sill plates hide about an inch of carpet, you want the carpet to go as close to the outer edge of the sill plate as possible. It is fine to save trimming the sill plate areas until after the seats are installed. This way you know how much carpet will be compressed or bunched up under the seats.

When you are satisfied with the trial fitting, remove the carpet and, depending on the method you have chosen, either prepare it for gluing or attach Velcro or tape. Most cars have a front and a rear piece of carpet. Install the rear section first since the front overlaps (under the seats) the rear carpet. If you are gluing, only apply glue to one half of the carpet at a time. Fold it back at the transmission tunnel and spray both the bottom side of the carpet and the top side of the padding with contact cement or trim adhesive. Carefully fold the carpet from the center to the outside edge, using your hands to smooth out any wrinkles. When the first side is in place, fold up the other side and repeat the process.

Front Section—With the front section, attach the driver's side first. There are more items to fit the carpet around, so this side is more critical. Remember to install the dimmer switch grommet if one is used.

When the carpet is secured and you are satisfied with the fit, vacuum the carpet to remove any possible installation debris. It might seem excessive to vacuum a brand new carpet, but the cleaner you keep your carpet, the longer it will last. Any debris will act like an abrasive and cause premature wear. An added preventative maintenance step that some installers recommend is an application of a fabric protectant such as 3M Scotch Guard fabric protectant.

If you were only replacing the carpet, reinstall the seats and accessories. Be careful that the carpet filaments don't get bound up in the bolt threads. If this starts to happen, you will need to enlarge the carpet opening slightly. ∎

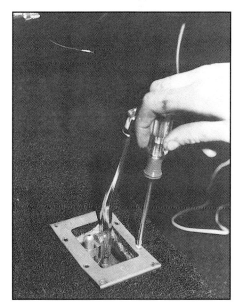

An awl is helpful in locating the screw holes for items like the shifter plate. Shifters can get sloppy from years of use. Now is a good time to install a shifter bushing repair kit so the unit will work like new. Reproduction outer trim plates and shifter boots are available.

49

CARPET

If the fasteners for your seats are reached through the bottom of the floorpan, chances are very good that the rubber plugs are either damaged or missing. Installing new plugs will help keep water out of the floorpan.

Cars like station wagons are usually carpeted on the back side of the rear seat and in the cargo area. Pre-cut carpets aren't always available for this part of the car, but you can get extra matching carpeting in a roll. An easy pattern can be made by cutting newspaper to fit the area. Tape the pieces of paper together when you are satisfied with the pattern.

Using a big, flat work space, place the pattern on the unrolled carpet and mark the outline with chalk. Then cut out the carpet with heavy-duty scissors.

Use trim adhesive to secure the new carpet to the seat backs and cargo floor. This technique can be used for other vehicles like vans or pickup beds.

CARPET

This is the same car as in the first caption. A C.A.R.S. reproduction Daytona weave (just like the factory original carpeting) carpet kit was installed over a Roadkill carpet pad. New pedal pads and door sill trim plates were also installed.

If you are fortunate enough to have a collectible car like this Camaro convertible with the rare fold-down rear seat option, the seat back carpet should be replaced when the floor carpeting is changed. Not all carpet kits come with provisions for a fold-down rear seat, so be sure to ask when you purchase the kit.

Not all vehicle floors are covered with carpeting. Some more basic models (and lots of older trucks) have rubber floor mats. The unique Chevy Nomad station wagons use this heavy, ribbed cargo deck material, which is available from C.A.R.S.

This Pro Street style mini truck is a good example of how carpeting can be used in a custom mode. Indoor/outdoor carpet is best for truck beds unless the truck has a canopy.

Door Panels

6

One of the easier upholstery tasks is door panel replacement. If you are fortunate enough to have a popular car or truck with a large supply of replacement and reproduction parts, upgrading your door panels is basically a simple remove and replace operation. If you have a vehicle without reproduction door panels, there are still things you can do to repair the panels and/or replace them.

Door panels are easy to access, but it is a good idea to check a factory shop manual to determine just how the panels are attached to the door. A variety of different fasteners are used to secure the door panels. Some of these fasteners are hidden behind things like the armrests. Many door panels involve some type of a clip retention system. Not all of these clips are reusable. Be sure that replacement clips are readily available before you force out the old clips. A detailed factory trim manual will provide exploded views of the doors to show

Sometimes door panels can look OK from a distance. This Corvette door panel seems acceptable in a black and white photograph, but actually it was quite faded and cracked in several places. The door panel is molded plastic so a whole new reproduction panel is required. Notice the plastic covering on the exposed inner door. If it is ripped, it should be replaced.

Door Panels

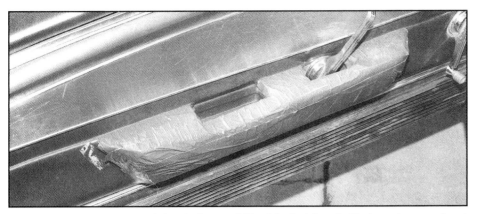

Other door panels are more obviously damaged like this duct taped Mustang door panel and armrest. This door panel has many different and complex elements such as stainless inserts and horizontal ribs.

Power windows were relatively rare on fifties cars like this '55 Chevy Bel Air. The backing board behind the new reproduction door panel has perforations for both power and manually operated windows. The buyer opens the holes for his particular application.

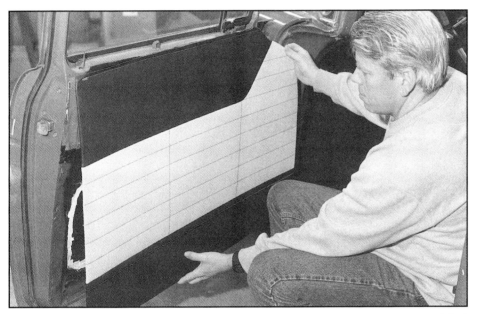

The reproduction door panels come pre-sewn, but without any trim. If you have good original trim, you can save money by reusing it. There are no holes in the panel when it comes from the manufacturer.

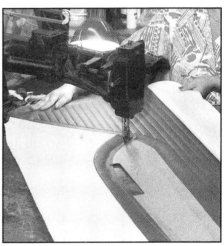

Some reproduction door panels are made of several different types and colors of vinyl. Manufacturing all these elements is expensive, which is why the more complicated door panels can be expensive. The material is sewn to the sturdy backing board with an industrial sewing machine using the correct thread and style of stitching.

you where and how the door panels are secured.

Most vehicles have door panels that cover the entire door, but many trucks and some cars (like early Mustangs) have partial panels. The partial panels are surrounded by the metal facing of the door. The paint on this part of the door is frequently chipped or scratched. You may need to touch up these areas or you might want to change the color. If you want to change the color, consider using a paint stripper to remove the old paint. If the metal part of the door is textured (like on Mustangs) be careful not to sand off these texture lines. Refer to the chapter on painting metal parts for more details on door painting.

DOOR PANEL REMOVAL

Whether or not you have a shop manual, the first place to start for removing the old door panels is with the armrests, door handles, and window cranks. The armrests are usually secured with several Phillips-head screws. Sometimes these screws are down inside the recessed part of the armrest. Keep track of which screws belong in which holes.

Door Panels

Door handles and window cranks that are secured with retainer clips need a clip removal tool to get at the hidden clips. The open part of the clip is supposed to face the long part of the handle, so this is how the removal tool should be positioned. Some handles use screws instead of clips.

This early Mustang uses a door panel insert instead of a full door panel. These door panels are much less expensive than full panels. A door panel clip removal tool is inserted along the edge of the panel where the spring clips are located. Gentle prying will release the clips.

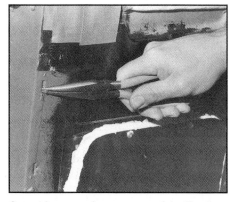

Some door panels are secured to the door with small nails or tacks instead of clips. If the panels were removed in a careless manner, there may be tacks left behind. The old tacks need to be removed before the new panels can be installed.

Handles & Cranks

Door handles and window cranks are often held in place by small horseshoe-shaped wire clips. There are inexpensive removal tools that slip between the handles and the door panel. The tool engages the ends of the clip and you push it off the slots in the handle. The handle is splined so you might have to wiggle it a little to remove it. If the handle doesn't come off easily, you probably didn't release the retention clip. Some door handles are held in place by a screw in the center of the handle. The screws are usually Phillips head, but you may also find Allen head and Torx fasteners. This screw can be hidden by a glued-on section of chrome trim. If you have the glued-on handle centers, remove them slowly and carefully so you have the best chance of reusing them.

Spacers—When you remove the door handles and window cranks, you will usually encounter circular plastic spacers which go between the handles and the door panel. These spacers help protect the upholstery. Underneath the door panels you will find springs which put the proper tension on the handles to keep them from rattling. Be sure to keep the springs and put them back when you reinstall the door panels.

Moldings—Some styles of door panels are partially retained by the window garnish moldings. These moldings are usually fastened with Phillips-head trim screws, but they can also have mounting clips. If the moldings don't fall off in your hands, try prying gently to remove them from the clips. Remember to unscrew the door lock buttons before you attempt to remove the window moldings.

Some door panels are secured by trim screws around the perimeter of the panel. These are the easiest to figure out and remove.

Retaining Clips

Most door panels use some type of metal, nylon, or plastic retention clip. The plastic clips have serrations that hold the clips in the holes. These clips are almost always located in the corners, plus several other spots along the edge of the panel. Use a flat bladed upholstery prying/tucking tool to lift the edges between the door panel and the door. Pry at the shoulder of the serrated clip or right where the metal clips enter the door. If you pry elsewhere, you run the risk of damaging the panel board. There are

Door Panels

When the door panels are removed, take notice of how the handle springs are positioned. Usually, the smaller end is against the door with the larger end against the door panel. These springs are often rusty or missing, but new ones are readily available.

Giving the doors a thorough cleaning is a good idea at this point. The access plates are usually held on with a few screws and sealant around the panel perimeter.

The cover plates for the '56 Chevy door access holes were in sad shape. They had a lot of surface rust and old undercoating on them. My friend, Brian Kennedy, cleaned them up in his bead blaster. A wire brush and a lot of elbow grease will also do the job. I applied POR-15 silver paint on the insides of the cover plates and POR-15 black to the exterior of the plates and the main part of the door. Using a rust inhibiting paint now will help prevent future problems.

also special door panel pliers which are designed to pry out the clips. You can use a screwdriver, but exercise caution so that you don't damage the door panel (unless you are positive that you are going to get replacement panels and have absolutely no possible use for the old panels or clips). If you want to reuse the old panels, remember that the base of the door panels is just panel board (like a thicker version of poster board) and it can rip or bend easily. It is quite easy to rip the clips out of the board. When this happens, it is difficult to secure the door panel when you reinstall it.

It is possible to find some door panels that were glued in place. These panels can be quite difficult to remove without destroying the panel board base. It is also common for the door panels to have some type of metal or molded plastic top section. These sections are often pressed down over the top inside edge of the door. You need to remove the lower and side clips first and then gently lift or pry upwards to release the top section of the door panel.

Older vehicles like those built in the fifties often used upholstery nails to secure the door panels. Sometimes these nails went into holes in the doors and other times they were fastened to tack strips along the edge of the panels.

Power Controls

If your vehicle is equipped with items such as remote control mirrors, power windows, and/or power door locks, you will need to separate the controls from the door panel. The electrical components like power window controls usually have plug-in connectors. The adjustable mirror controls are usually a flexible cable. Often a set screw needs to be loosened or the cable needs to be separated from the bezel surrounding the joy stick. These items can be a little difficult to work on since the amount of freeplay may be just a few inches. If you have any doubts about how accessories are connected to the door panel, consult your shop manual.

Rear Quarter Panels

Depending on the vehicle design, the rear quarter panels can be similar to the front door panels. There is more similarity between front and rear in the older vehicles. Besides clips and screws, the rear quarter panels can be secured by fasteners hidden by the rear seat. The window cranks are secured in the same manner as the front cranks.

INSIDE THE DOORS

Water Shield

When you remove the door panels you will probably see some type of insulation or water shield on the door skin. These paper or plastic coverings are frequently ripped or otherwise damaged. You can purchase replacement water shields or make your own out of similar thickness paper or plastic. The water shield usually has some type of sealant

Door Panels

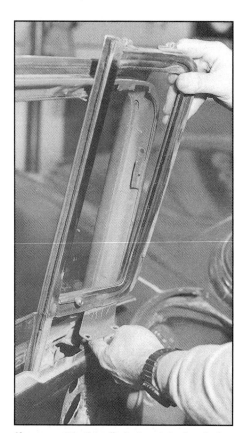

If you want to do a first class job of restoring doors, install new rubber seals and window channel whiskers. Remove the window garnish molding to gain access to the vent window fasteners. The vent pivot post needs to be unfastened from the crank mechanism so that the entire vent window assembly, including the division bar channel, can be lifted out of the door. The crank mechanism can be removed through the lower access opening. Then carefully lift the glass out of the door.

Position the window so that the four screws that secure the metal glass retainer channel to the regulator arms can be removed via the access holes. Then carefully lift the glass out of the door. If your side glass is chipped or starting to get foggy around the edges, go to your local glass shop for new glass. This is a perfect time to switch to tinted glass.

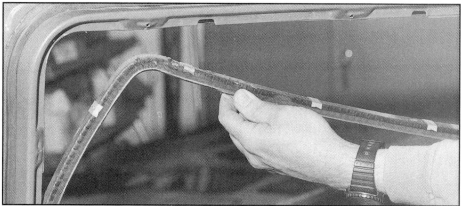

The original flexible channel is held in place with clips, so it is easy to remove. Start at the front of the door where the channel meets the division bar. The lower end of the channel is secured near the bottom of the door in a metal channel which is fastened with a stud and nut.

around the edges to secure the shield and keep out water. Remove the sealant carefully so as not to damage the shield.

If the old water shield is in good condition, you can probably reuse it. If it isn't in great shape, try to preserve its shape so you can use it for a pattern. It is always best to remove old parts carefully for this reason just in case there isn't a readily available replacement part.

Many people are tempted to skip dealing with the water shield, but remember that the shield helps protect the porous panel board backing of the door panels from warping. Rain runs down your windows to the inside of the door. Hopefully, your lower door edge drain holes are clear and working, but moisture can accumulate. The water shield is what keeps that moisture away from the porous part of the door panels. The shield also keeps out dust and performs a minor insulation function.

Inspection & Reconditioning

With the old water shield out of the way, take a flashlight or trouble light and inspect the insides of the door. Check to see that the drain holes are not clogged, and look for serious rust spots. Put the window cranks and door handles temporarily back in place so you can check the operation of the cranking mechanisms and door lock components. Now is a good time to lubricate the moving parts. White lithium grease works well as a lubricant. Put it on all the wear points. Some of these inner parts are secured with clips. Make sure that the clips are snug. The lock assemblies often wear

Door Panels

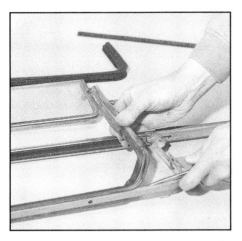

The vent window must be completely disassembled in order to install the new C.A.R.S. gasket set and division bar fuzzy channels. The glass frame has a small post on the top of it which must be carefully pried out of the outer framework. Remove the old vent rubber seal. The longer section is a press fit, but the vertical seal has retainer tabs.

In order to install the new whiskers on the metal window molding, you need to position and clamp the new whiskers on the garnish molding so that the stainless steel bead is even with the top edge of the molding. Then use the steel template (supplied with the kit) and a 3/32-inch drill bit to drill through both the whiskers and the garnish molding. Drill the holes approximately every six inches.

The new whiskers are secured with small staples. Push the supplied staples through the holes in the whiskers and the garnish molding. The stapes are inserted from the whisker side. Use needle nose pliers to bend the staple legs to the inside of the garnish molding. Squeeze the pliers tightly so that the staple is hidden down inside the whiskers for a secure fit.

out on older cars. While you have the door apart is a great time to replace defective or marginal parts.

You may even want to fine tune the fit of the glass in the door. Consult the shop manual to see if and how your glass is adjustable. Be careful not to drop or otherwise damage the glass while you are adjusting it.

Window Felt

Now is a great time to inspect and replace the window felt or whiskers. These are the strips that go along the edge of the windows to help keep out moisture. These strips are usually attached with clips and/or screws. While you are doing all this fine tuning of the doors, you might as well install new door edge weatherstripping. If you plan to repaint the vehicle, wait until after you have painted it before replacing the door weatherstripping (but remove the old weatherstripping before painting for a better paint job).

In order to make the restored interior as quiet as possible, I added insulation to the inside of the doors. I installed Stinger Pro Roadkill sound damping material in the doors and quarter panels. I cut 38x24-inch pieces for the front doors.

Insulation

While the door panels are off, consider adding an extra layer of sound deadening insulation. There are many high tech insulation materials available, and the best place to look

I used Pro Roadkill because it has its own adhesive and doesn't require the use of a heat gun. I inserted the insulation material through the window opening and put my hands through the access holes to press the Pro Roadkill against the door. An awl was used to open up the side trim holes in the insulation. By opening the holes from inside the door, there is less chance of pushing the insulation away from the door when the stainless trim is installed.

are high end car stereo installers, who often use this material to pack the doors when installing mega-watt

Door Panels

Before the access hole cover plates are installed, it is a good idea to spread a bead of all purpose body seam sealer around the openings. The goal is to seal the inner door as tightly as possible.

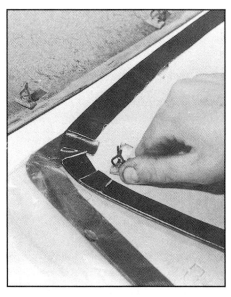

Some door panels don't come with new clips (although you can get them) so the old ones can be reused if they are still in good condition. Don't throw away any of the original parts until after the whole restoration project is completed. Here, the old panel was checked to see which way the clips should be mounted.

Here is the restored and reassembled door all buttoned up. All the inner door mechanisms were checked and lubricated. Small miscellaneous holes were covered with duct tape. Sealing the door reduces noise transference and protects the cardboard back of the door upholstery panels from moisture.

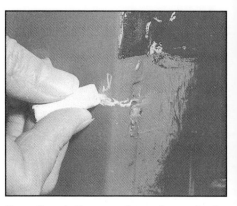

Even though the mounting nails are in the correct position in the C.A.R.S. side panels, the holes can be marked with a piece of chalk. This makes it easier to line up the nails without bending them or scratching the door paint.

stereo systems. Some come in rolls and others are available in sheets. Some have their own adhesive and others require that you apply them with a quality trim adhesive or glue. Anything you can do to lessen the drum-like characteristics of your car will make it quieter. Many older trucks had virtually no factory insulation so they are excellent candidates for aftermarket insulation.

The insulation materials come in various thicknesses. You want a product that is designed for doors. The product needs to be thin enough so that it won't make the door panels bulge out.

Cover as much of the door as possible. Make the openings for door handles and window cranks as small as you can. Besides absorbing sound, the insulation material blocks sound.

Rust Proofing

You may want to apply some brush-on or spray rust-proofing material to the insides of the doors. Be sure that the glass is out of the way when you apply the protectant. Most spray rust inhibitors come with extension tubes to make reaching the far corners easier. The excess rust-proofing material may seep out of the lower drain holes, which should be wiped off with a solvent such as mineral spirits. You can also tape over the drain holes (from the outside) with duct tape before applying the rust-proofing material. Remember to remove the duct tape after the rust proofing has dried.

DOOR PANEL RECONDITIONING OR REPLACEMENT

While the door and quarter panels are removed from the car, you can carefully inspect them to determine if they need to be replaced or just touched up. Pay close attention to the bottoms and edges of the door panels. These are the areas that usually get damaged. Lower edges can get water damaged or scratched, scuffed, and/or ripped from years of hard use.

Replacement

The primary determination concerning the replace or repair issue

Door Panels

is whether reproduction door panels are available for your vehicle. Collect several catalogs from restoration sources before you tear your car apart. If you have a car that is too new to be serviced by the collectible car industry, consult the parts department of your local new car dealership that carries your brand of car. They might not have replacement door panels in stock, but they may be able to order them for you. Cooperative parts departments can sometimes do computer searches of other dealers' stock for parts numbers that aren't current. You may also be able to find sources for slightly out of production interior parts in *Hemmings Motor News*.

On vehicles that aren't too old, you may be able to find better door panels in a wrecking yard. You want to find a vehicle that ended up at the junkyard because it was wrecked, not because it was abused and worn out. Finding excellent door panels in a wrecking yard could take some searching.

Occasionally you may have trouble finding the same color reproduction door panels if your vehicle has an unusual interior color. For obvious reasons, manufacturers start with the most popular colors. They may or may not expand the colors depending on the market demand. If you can't find your exact color, you can get one of the available colors and have the panels dyed to match your interior. More information on dying upholstery is in the chapter on color changes.

Reconditioning

If your door panels are in really good condition, they may only need a thorough cleaning. It is easier to clean all the little hard-to-reach areas when the panels are off the doors and the armrests have been removed. An old toothbrush works well for getting dirt out of the seams and crevices. Be careful not to use a harsh cleaner or too much elbow grease. The stitched seams can be damaged.

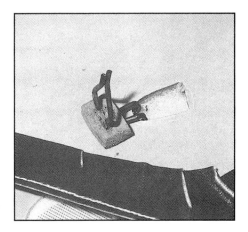

This close-up of a retaining clip shows how it is hooked on the backing board through the small hole in the panel.

After cleaning the door panels there may be some noticeable scratches or areas that are discolored. You may want to use vinyl paint to repaint the panels in their original color. Using the original color gives the best results with vinyl dye and should make the door panels almost as good as new.

If your door panels have some relatively minor damage, such as a ripped lower section, you might be able to save money by having an upholstery shop repair just that section. Get an estimate and compare the repair costs to that of new reproduction door panels. An upholstery shop could also recover the armrests if they are the only worn parts on your panels.

Molded Door Panels—Some classic cars used molded door panels. These panels are like a big dashpad, with foam padding underneath the vinyl covering. This style of door panel is prone to cracking. Fortunately, there are specialty firms

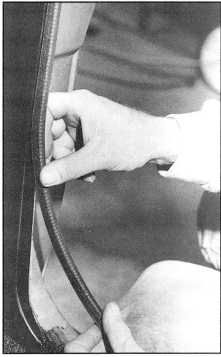

Some door panels or side panels (like this Mustang rear quarter panel) are finished with a flexible molding strip or plastic wind lace. They help secure the panel to the door jamb and add a finishing touch.

that can restore and recover damaged foam-padded molded door panels and dashboards. You can find these companies in enthusiast magazines and *Hemmings Motor News*.

Many door panels use some type of stainless steel trim or ornamentation. Depending on your vehicle, these trim items may or may not be available. You should save the trim items until you are sure you can get new replacements. If the trim is in good condition you can reuse it.

Check the back side of the door panels to see how the trim is retained. Often there are little tabs that go through the panel and are bent over. These tabs can be fragile so exercise care when removing them. The stainless trim can be brightened up by polishing it with metal polish. If there are any heavily oxidized areas, gently rub the trim with a piece of fine

Door Panels

Here is the Mustang door that was repainted in Chapter 3 with the addition of reproduction door panels, new armrest bases and pads, and the original handles, which were cleaned and polished. It is a major improvement from the scratched and dirty original door.

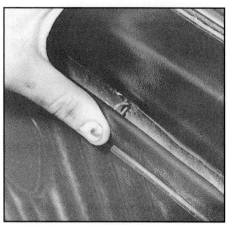

When the new door panel is about to be installed, look to see that the retaining clips are aligned with the corresponding holes in the door. The clips can be moved easily so that they line up with the holes. When the clips are in the right position, push the panel against the door.

bronze wool. Then polish the trim.

If the trim has somehow been dented, you can remove and straighten minor dents. Use a small trim hammer and back the area you are working on with a soft piece of wood. You can also shape the wood to fit inside the back of the trim and use the hammer to tap on the front side. A little judicious filing may be necessary to remove any high spots. After a dent has been removed, you will probably need to polish the trim in steps with ever finer compounds. This technique is the same one restorers use to repair outside body trim, but the interior trim pieces are more fragile.

INSTALLING NEW DOOR PANELS

The installation of new reproduction door panels is pretty much a reversal of how you removed the original panels. Depending on how complete your reproduction door panels are, you may need to reinstall items such as the trim, armrests, and retention clips. Always save the old door panels until you have installed the new ones. You may need to refer to the old panels for things like the locations of the clips.

Many reproduction door panels have perforations on the back side for the various handles and accessories. Check your old panels to be sure that you only open up the holes that match your car. The underlying panel board could easily have the perforations for power window switches when your car doesn't have power windows.

If your door panels don't have precut openings for things like window cranks, check carefully before doing any cutting. The old adage about checking twice and cutting once definitely applies here. You can position the new panel on the door and feel where the handle shafts protrude. Make a very small slice to release the shaft. You can always make a small hole bigger, but you can't do the opposite. Make sure that any holes are small enough to be covered by the handle base or trim escutcheon.

If your car or truck has the escutcheons around the door handles (the recessed plastic areas that fit around and behind the handles) they are often deteriorated from years of use. Your old ones will look extra shabby against the new door panels. Some of these escutcheons are still available from the dealership, while others must be purchased from reproduction parts companies.

Your new door panels might not come with the little grommets that go around the door lock stems. You should install new grommets for a factory-finished look to the door panels. These grommets can be found at upholstery supply stores and even some regular auto parts stores. Many auto parts stores have a section with blister pack interior parts like door

Door Panels

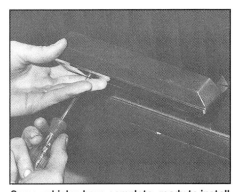

Some vehicles have complete, ready-to-install armrests available. This is the easy way to restore the armrests. In other cases, the old base needs to be used with pads that you recover.

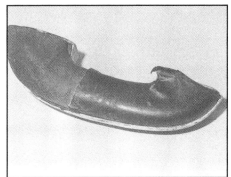

Old armrests, especially the driver's side ones, can take a lot of abuse. It is common to see taped up armrests. Some armrest bases are reproduced, but in many instances, the original base must be reused.

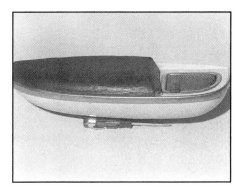

Some armrests are far more difficult to find than you would imagine. Classic Chevy rear seat armrests are especially hard to find if you don't have the original ones. The ashtrays tend to get rusty. The small spring that is attached to the lid is often broken or missing.

Here is a disassembled '56 Chevy front armrest. The elements are (from left to right) the stainless trim, the plastic base, the foam pad with its metal base plate, and the old vinyl cover.

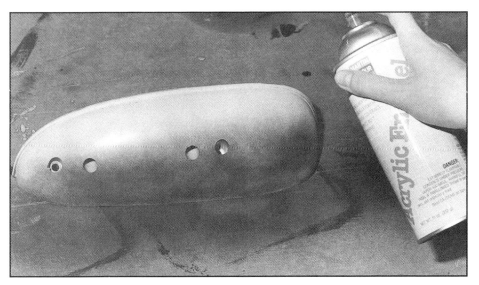

The armrest bases should be cleaned in a solution of water and Simple Green. The bases need to be free of dirt or grease so the new paint will adhere properly. My used armrest bases were solid and free of cracks, but they were three different colors so I painted them all black. The chrome on the ashtray lids was rusted so I bead blasted them and painted them black also.

handles, window cranks, and retention clips.

Armrests

Armrests can be a problem if you don't have them. Not all armrest bases are reproduced, so check before you toss your old ones. Recovering the actual padded and upholstered section isn't a problem, but you need the base. If you find a good base that is the wrong color, you can change the color with spray dye.

If your armrest base has a chrome plastic finish, this finish often gets tarnished and worn. There are specialty companies that can rechrome plastic. You will most likely have to send away for this service.

Recovering—Many door panel kits give you some extra matching material for recovering the armrests. You may need to replace the foam padding if it is too worn. You can use the vinyl covering of the old armrest as a pattern for the new covering. Contact cement will secure the vinyl to the foam. The biggest thing to watch out for is wrinkles. Make sure that the vinyl is room temperature before you install it.

Door Panels

Contact cement should be applied to all mating surfaces of the base and the vinyl. You can usually tell by looking at the old cover where the cement should be applied. Remember, contact cement needs to dry before the two surfaces are joined.

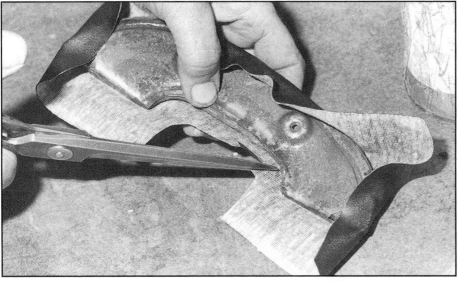

With the ends secured, you can start trimming the excess material with a scissors. After the basic shape has been established, cut triangular darts to help the vinyl lay flat in the curved areas.

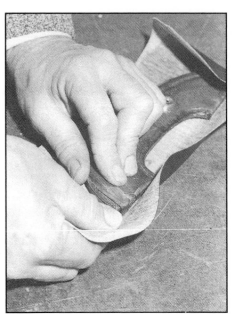

For this front armrest I left the vinyl uncut and wrapped the excess material over the ends of the armrest base and foam pad. Secure one end first and then stretch the vinyl over the opposite end. Pull the vinyl as taut as possible to avoid wrinkles.

The door panels usually need to be positioned at the top of the door first. This is a must if the panel fits over or clips to the top of the door. You want to insert the retention clips with as little force as possible so that you don't damage the upholstery. If you have the clips properly aligned with the holes, you should be able to insert them with just the pressure of your hand. If you think you need a hammer, check to see that there aren't any obstructions. Remember to connect any electrical fittings before you secure the door panels.

If for some reason you either can't find the correct door panel clips or if you have a section of the panel that just won't lay flat, you can use hook and loop fabric strips to secure the panel. Glue the strips so that they line up with each other. Some hook and loop material comes with adhesive backing. The material can be found at most fabric stores.

Remember to install the window crank springs on the inside of the door panels. Place the handle washers on the outside before you add the handles. If your handles use the little horseshoe-style retention clips, push them all the way on before you install the handles. When you push the handle down on the splined shaft, the clips will snap in place. Take care when you position the handles. Make them uniform throughout the car. Reinstall the window garnish moldings if they are a part of the door panel retention system.

NO-SEW CUSTOM DOOR PANELS

The focus of this book is easy, "no sewing required" interior improvements. Most door panels require

DOOR PANELS

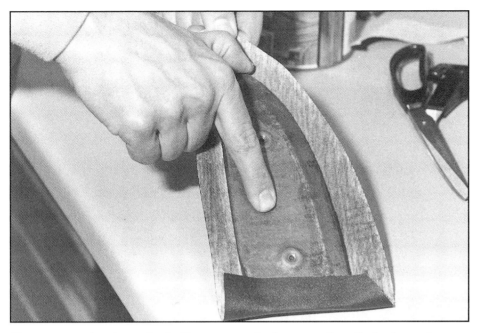

The rear armrest pads on classic Chevys are easier to recover than the front ones because their shape is less complex. Start by attaching the vinyl to the flat, wide end. Then pull the vinyl toward the tapered end and secure it to the metal base.

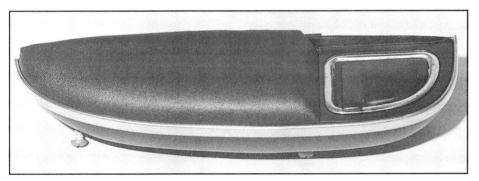

Here is a restored rear armrest. I had to paint the ashtray lid black since the chrome was shot. If I ever find better ashtrays (or if anyone ever reproduces them) I can easily upgrade. These armrests are a great improvement over their worn out starting condition.

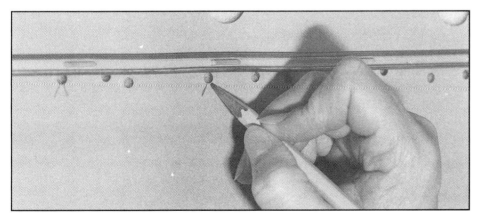

All the factory stainless interior side trim was missing from my '56 Chevy project car. Fortunately, I was able to obtain reproduction pieces from C.A.R.S. The panel boards used behind the vinyl are somewhat universal. They can be used for more than one model of '55-'57 Chevy so there will be extraneous trim mounting holes. Take the stainless and carefully locate and mark the holes for your application. Be careful not to open any incorrect holes.

The stainless trim pieces on the armrest bases can be shined up like new with metal polish. The armrest trim is held in place by the ends of the trim which are bent around the corners of the plastic base. The trim needs to be pulled away from the base slightly to reinstall it. Be careful not to bend or kink the trim.

sewing or the molding of vinyl with special equipment, but it is possible to make some basic custom door panels without sewing. We will briefly discuss two similar styles of custom door panels: flat-glued door panels and sculptured door panels.

These types of custom door panels work best on the most simple doors like those found in older cars or trucks. They also work well on vehicles that have partially upholstered doors. It is easiest when the panel is flat.

Flat-Glued Panels

The basic idea is to cut a piece of fabric or upholstery material and glue it to the panel. The edges of the fabric are tucked around the back side of the panel and glued or stapled in place. Holes are cut for door handles and window cranks.

Cutting Panel Board—If you have a utilitarian vehicle like an older truck that didn't originally come with door panels, you will need to make a pattern and use upholstery-grade

Door Panels

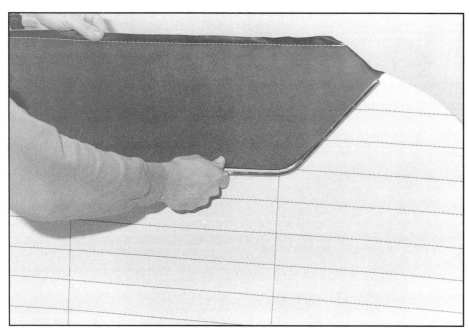

The stainless trim fits right between the black and white sections of these door panels. The mounting tabs are small so use care when pushing them into the door panel.

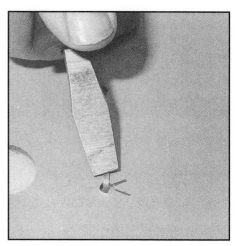

Always try to make the smallest possible hole in the vinyl (you can always enlarge a hole, but you can't shrink it). I used an awl to dimple the vinyl from the back side just enough to be seen on the front side. Then I made the actual hole from the front side of the panel. I used the tip of a flat bladed screwdriver to carefully bend the trim mounting tab against the back side of the door panel.

Use a small X-Acto knife to make "X" openings in the vinyl for the window cranks and the door handles. Start with small openings and enlarge them as necessary.

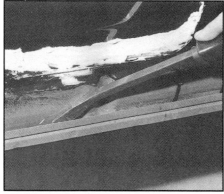

There is a side panel retaining strip that runs along the bottom edge of the front doors on classic Chevys. This one was crimped in several places so I used a small pry bar to open it up. There is a metal strip on the bottom edge of the side panels that fits inside this retaining strip.

waterproof panel board for the basis of the door panel. If you build a panel, undercut it slightly to leave room for the thickness of the material and a little general clearance. Undercutting about 1/8 inch usually works well.

Drill Holes—When you have the panel board cut to the proper dimensions, you will need to drill mounting holes (if the vehicle has never had door panels before). Get the clips (from an upholstery supply store) first and match the holes to the clips. Use a stop on the drill bit so that you don't go too deep and damage anything inside the door. If you are recovering an existing door panel, the job will be much easier.

Padding—A thin layer of upholstery foam can be glued to the panel board to give the door panel some padding and thickness. Or, you could just glue the fabric to the bare panel. Most people opt for the foam padding.

Cover—For a plain, flat-glued door panel, choose your covering and cut it at least one inch bigger all around than the pattern. If you are using foam, it is probably a good idea to make the material 1-1/2 to 2 inches bigger. You can always trim the excess material after it is bonded to the panel board.

Use contact cement to bond the upholstery material to the panel board or foam. Follow the glue directions and apply the contact cement to both surfaces. Be careful when you mate the two surfaces. Position the fabric so you don't need to readjust it. Start in the center and smooth out the fabric as you push it tightly against the mounting board or foam padding.

The back side of the material will need to be notched in order to smoothly blend around the corners. You don't need a lot of notches. Make

Door Panels

With the front door panels, start at the bottom edge and engage the retaining strip (if the vehicle has one) that runs along the bottom of the door. Then carefully locate the mounting nails (or clips) in their respective door holes.

When the nails are in position, use your fist to tap the mounting nails all the way in. You can use a rubber mallet, but be sure that whatever you use doesn't mar the vinyl.

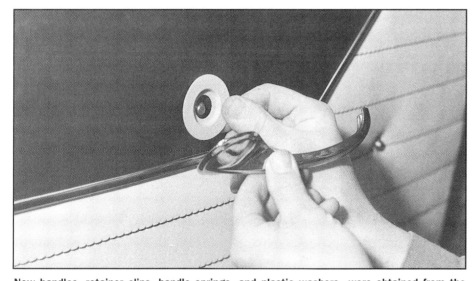

New handles, retainer clips, handle springs, and plastic washers were obtained from the C.A.R.S. catalog. The plastic crank washer goes between the door panel and the handle or crank to protect the vinyl and make the handle operation smoother. I installed the crank retaining clip on the handle before pressing the handle on the splines. The clip should snap in place if you press firmly on the handle.

sure that the cuts stop short of the edge of the mounting panel. The fabric can either be glued or stapled to the back side of the panel. If you use staples, be sure that they are short enough that they don't protrude through the front of the door panel.

A variation on the basic, no-sew door panel would be to include a strip of carpeting along the bottom six inches or so of the panel. The carpet could be glued to the panel or secured with hook and loop strips. The carpet would look more finished if it had a binding around the edge. You could cut and fit the carpet and then take the pieces to an upholstery shop for the binding.

Sculptured Panels

The idea behind the sculptured door panels is basically the same as the flat panels except that more foam is added in some places and foam is removed in others. The fabric (often tweed or burlap style upholstery material) is glued to the foam and you have that high tech sculptured look.

A relatively dense foam needs to be used for sculptured panels. A high density sealed surface foam known as white landau foam works well. It's about 3mm to 4mm thick.

Most people make either simple ridges or geometric designs in their sculptured door panels. Experienced

Door Panels

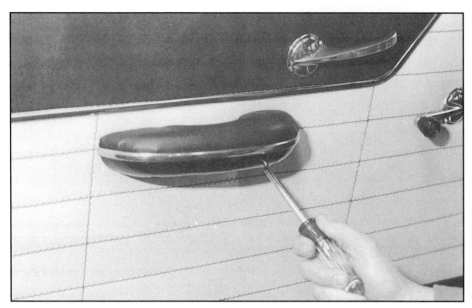

Two large screws secure each armrest to the side panels. The screws are angled sharply so it may take a little poking around to find the screw holes.

Here is the completely restored '56 Chevy front door panel. It looks factory fresh. Since the original panels were completely missing, the doors could not have been restored so easily without the availability of so many great reproduction parts.

The window garnish moldings need to be installed since they cover and secure the top edge of the side panels. I used all new mounting screws from C.A.R.S.

design is, you should probably allow three inches of excess fabric. This way you will have plenty to fit into the sculptured sections and still be able to wrap the edges around the base board.

Use contact cement to secure the fabric to the foam. Start in the center of the panel and work your way out to the edges. You don't want any material to bunch up. The fabric needs to be gently pressed into the grooves of the design. You want the shape to be well defined, but you don't want to rip the material. An upholstery tucking tool will work well to get the fabric into the sculptured areas.

The techniques used for glue-on door panels can also be used in trunks and for simple headliners. Many older trucks and the less expensive models didn't come with headliners. This technique will work well for making a basic rectangular headliner that just covers the flat part of the roof. You could glue the fabric to the sides of the cab, but this would be much more involved. Kick panels and a rear package tray could also be made with panel board, foam, glue, and fabric without any sewing. ■

upholstery workers can make flames or emblems such as the Chevy Bowtie. The simple designs are best for beginners.

You can cover the entire door panel in foam and then add thicker sections for the sculpted look, or you can cut away sections of the foam so that the base is exposed. It all depends on how thick or intricate you want the panel to be.

Depending on how elaborate your

DASHBOARDS 7

Restoring the dashboard is something many people ignore, but it isn't particularly difficult. If you restore the rest of your interior, a shabby dashboard will detract from all your other work. You look at the dashboard, gauges and related components every time you drive the vehicle, so why not improve the scenery?

Dashboard restoration falls into four general categories: non-padded dashboards, padded dashboards, cleaning gauges and trim, and sheetmetal repairs. As with most interior jobs, the ease of the repairs often depends on how your vehicle was constructed and how easy it is to find replacement parts. You are most likely to find simple, uncluttered dashboards with minimal gauges and controls on earlier model cars and trucks. Truck dashboards stayed basic longer than most cars. Dashboards that are either all metal or metal with a removable padded dash top are the easiest to work on. The more complicated, molded plastic and vinyl dashboards usually require the replacement of large components. When the dashboard can be separated into smaller components, it is much easier to work on.

As usual, I strongly recommend that you get a detailed factory shop manual with a complete interior trim section. Sometimes the body and trim information is in a separate volume from the mechanical and chassis shop manual. The type of repair manuals found in auto parts stores are condensed versions of the factory shop manuals. These books are fine for a general overview, but most of them seem to skip most of the interior information. The factory shop manuals will provide larger illustrations so you can see how dashboard

The dashboard of your vehicle is something you look at every time you drive, so it makes sense to restore it while the rest of the interior is being upgraded. Reproduction companies make a lot of dashboard trim, knobs, and gauge lenses for popular cars like classic Chevys.

Dashboards

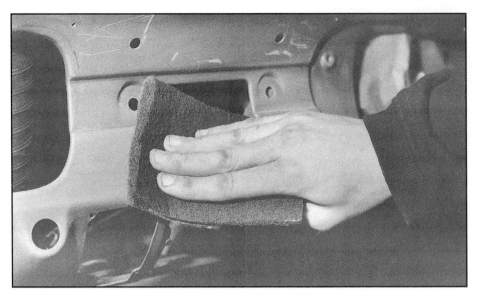

Unless the dashboard surface is badly scratched, you should be able to scuff it up for paint with a Scotchbrite® pad. Areas that have been scratched down to bare metal need to be featheredged (tapered out from the bare metal) with automotive sandpaper.

components are installed and secured. Fasteners can often be in unusual, remote locations, or hidden behind other components. If you know that there are still a couple fasteners to remove, you won't try to force a component that is still attached to the body.

NON-PADDED DASHBOARDS

Standard, non-padded dashboards are the most basic and easiest to restore. The basic needs are cosmetic, usually involving painting the whole dash or just touching up scratches. Scratched areas are those most used places such as near the keys, the radio, the ashtray, and the glove box.

Most people prefer to deal with the dashboard while it is still in the vehicle. For a superior restoration, you are better off to remove the dashboard from the vehicle. Removing the actual metal dashboard isn't as difficult as dealing with all the wiring components. Whenever you remove any wiring component, always label it with a piece of masking tape. Making drawings is a good idea, even if you have a detailed shop manual. A wiring diagram will tell you which color wires go where, but your own notes can be more meaningful to you.

You will find it easier to work on the dashboard if you remove the steering wheel. The steering wheel can be a real annoyance when you are trying to work on the left half of the dash. If you plan out your interior restoration, you can restore or replace the steering wheel at the same time as you are working on the dashboard, so removing it now makes perfect sense.

When painting the dashboard, it is best to remove as many components such as the gauges, trim, radio and switches. If you can't remove an item, be sure to thoroughly tape it. If you are changing the color of the dashboard, removing parts will give you a better job. If you are simply repainting the dashboard the same color, taping will be okay. An added benefit to removing dashboard components is that you can do a more thorough job of cleaning the parts. You would be surprised how much dirt and dust can accumulate over the years.

Painting & Repairing

Painting a metal dashboard is like painting any other metal interior part. Consult the chapter on painting interior components for specific painting details or pick up a copy of HPBooks' *Paint & Body Handbook*. A key element to remember when painting a dashboard is cleanliness. If you have ever used a silicone-based protectant on your interior, there can be silicone residue on the dashboard. Grease, silicone, and other contaminants can be the downfall of any good paint job. Be sure to super clean the dashboard before attempting to paint it if you want a quality paint job that will last a long time.

If your dashboard is in decent condition and the repaint is as much for freshening the old paint as redoing it, you can do a less thorough prep job. A good cleaning followed by a light scuffing with a Scotch-Brite® pad and then a wipe down with wax and grease remover should be sufficient for a same color repaint.

If you are changing colors or doing any repair work, the prep work will be more involved. Think of how you would repaint the exterior of your car; the dashboard techniques are similar. Repair work should be done first, but it isn't a bad idea to degrease the dash even before doing repairs. The less grease you have on the surface to be painted, the better the end results will be.

The type of repair work I am talking about is mostly dents or scratches that

DASHBOARDS

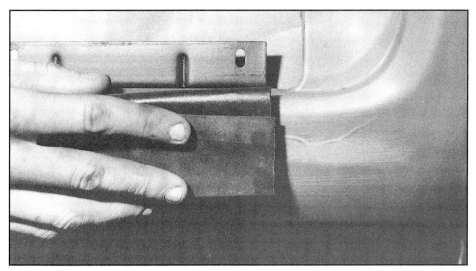

When you look closely at the dashboard, you might find drips or other flaws in the factory applied paint. Now is the time to remedy those flaws. There was a big paint sag under the glove box door so it was sanded down to the same level as the rest of the paint.

any paint project, the quality of the underlying repairs and prep work are the key to a good topcoat. Paint is a reflection of the underlying work.

PADDED DASHBOARDS

I will discuss three types of padded dashboards: the partially padded dash where the pad is molded vinyl over foam; the older stretched vinyl over foam partially padded dash; and the more modern single-piece, molded padded dashboard. Padded dashboards take a lot of abuse from the sun and the resulting heat inside a car. You may think those big fold-out window shades look silly, but they do a lot to keep temperatures down inside locked cars on a hot day. They also help protect padded dashes.

Partially Padded

The oldest style padded dashboard is the partially padded stretched vinyl. These dashboards were often optional on cars in the fifties and early sixties. Chances of finding a mint one in a car

are too deep to sand out. If some do-it-yourself stereo wizard owned your car, you may have to deal with a butchered radio opening. For popular cars like early Mustangs, there are patch panels for the radio area. The original Mustang radio opening was smaller than most aftermarket stereos, so many owners used a hacksaw or a file to enlarge the hole. These patch panels involve welding, so you may want to remove the dashboard from the car and take it to a body shop. You may find it easier to buy an uncut dashboard from a wrecking yard or at a swap meet.

If someone has kicked the dashboard or you have a couple deep scratches, you can probably do the repairs with the dash still in the vehicle. The determining factor is how accessible the dent is. Deep dents should be removed with a hammer and dolly. Dents and scratches of less than a quarter inch can be fixed with body filler. Be sure to featheredge the dent and apply enough filler so that it tapers out to the non-dented part of the dashboard. If you don't featheredge (the gradual tapering of the repair) the dent, you will have a noticeable line where the repair was made.

Scratches that are too deep to be just block-sanded out can benefit from an application or two of body putty. Body putty is essentially thick primer/filler that comes in a pliable tube. When it is dry, you block sand the area and apply primer over it. Like

Padded dashboards are prone to sun damage. There are many replacement dashpads available, plus there are companies that can restore and recover your original pad. Check your shop manual to see how the dashpad is secured. Very often, much of the dash trim will need to be removed.

Dashboards

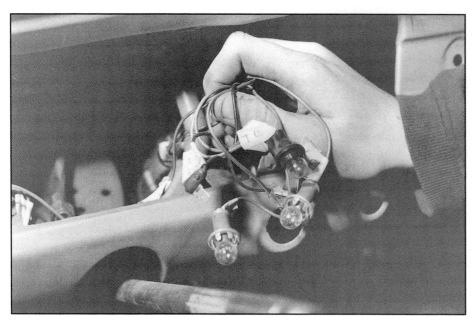

In the case of this Mustang, the dashpad color was being changed as were the metal portions of the dash. That required the complete disassembly of the gauges and accessories. All wires and bulbs should be tagged with a piece of masking tape so that returning them to their correct locations will be easy.

you buy are slim. Repairing such a padded dashboard is a task you may want to leave to a professional, but if you can find the correct vinyl, it is possible to do the job yourself. It requires removing the dashboard from the car.

You need to remove the old dashpad as carefully as possible so that you can use it for a pattern. The outer vinyl cover should be stretched over the underlying foam, but both the vinyl and the foam could be quite deteriorated. The foam could easily have shrunk, which will account for the wrinkles and voids in the vinyl pad. The metal part of the dashboard will need to be sanded so that all traces of the old glue are removed. If you sand down to bare metal, you will need to apply new primer and paint.

New Foam—Make a new foam covering out of high density foam. Hopefully you can use a single piece of foam for the pad. If you need to make any patches, try to put the patch seams in unobtrusive places or cover the seams with vinyl tape. You don't want the seams to show through underneath the vinyl covering. Use contact cement to glue the foam to the dash. After the foam is securely fastened to the dash, cut out any speaker or defroster duct openings.

New Vinyl—If you were able to remove the old vinyl cover in one piece, use it for a pattern. Otherwise, you will need to make a paper pattern and transfer it to the new vinyl. The old vinyl can be used to find matching material at an upholstery supply outlet. You may need to have the vinyl dyed to obtain the exact color your car needs.

The new vinyl should be a couple inches larger than the finished pad. You want extra material to wrap around the edges and for something to grip while you stretch the vinyl over the foam pad. Be sure that the vinyl is warm and pliable before you attempt to stretch it over the foam.

This type of padded dashboard often has some type of trim strip on the edge facing the interior. This is a good place to secure one edge. Then stretch the vinyl to the windshield side of the dash. You do not glue the vinyl cover to the foam, just to the front and

Some dashpads are glued to the metal portion of the dash and other ones are secured with small nuts which are often hidden under the dashboard. Don't force the pad up until you are sure that you didn't miss a fastener.

DASHBOARDS

The old dash foam can come apart and stick to the metal. Use a putty knife to remove the old foam or glue. The surface should be sanded so there is a uniformly clean surface for the new dashpad.

back edges of the dashboard.

Life will be much easier if you can get a friend to help with the next step. While one person pulls and stretches the vinyl into place, the other person can manipulate a heat gun (take care not to apply too much heat in any one place or you could damage the vinyl) or a hair dryer to work out wrinkles. You can also use the palm of your hand to smooth the vinyl toward the unsecured edge. When the cover is as smooth as possible, glue the back edge.

After the glue has properly set, you can trim off any excess vinyl material and reinstall any trim items. Reinstall the dashboard in the car.

Bolt-On Pad

The next type of padded dashboard is the bolt-on pad that was common in many sixties cars such as Mustangs and Camaros. The outer covering and the pad were a single unit with several mounting studs that match up to holes in the metal part of the dashboard.

Heat and sunlight damages the outer pads and the inner foam core often dries out and deteriorates. There are reproduction pads for the more popular vehicles. In these cases all you need to do is remove the old pad, call the mail order company, order a new pad, and wait for the UPS truck to deliver it.

If you have a vehicle that doesn't have any reproduction dashpads available, you can search wrecking yards for a better unit. You may have to settle for a different colored pad, but it can be dyed the correct color. If you can't find a suitable used pad, there are specialty companies that can restore your old pad. They will vacuum-form the vinyl to the pad, so the final result will look just like new. These services usually cost more than a reproduction pad (due to the custom nature of the process) but they are definitely reasonable. These dashpad restoration companies advertise in hobby magazines. You probably won't find such a firm locally, but the companies are set up to deal with mail order.

Minor Repairs—Padded dashes that aren't terribly damaged can be mended with dashboard repair kits. Restoration supply companies like The Eastwood Company offer these kits. Besides fixing small problems, these kits can help you get by until you can find a better pad or afford a new one.

The basic idea of the repair kits is to fill the cracked area with an epoxy

Here is a Mustang dashboard stripped down to the bare essentials. The steering column outer tube was removed, also.

Dashboards

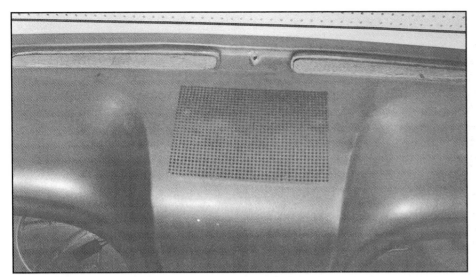

Many padded dashes have perforations in the center for a radio speaker. This area is very prone to cracking. Other weak areas are the leading edges of the pad.

Some reproduction dashpads are very expensive or not currently available. In those cases, you may need to settle for a used dashpad. The condition and color of the used pad will affect the price.

filler, sand it to match the shape of the dashpad, and repaint the area with the color compound included in the kit. The kits include the basic colors, which you need to mix and match for custom shades. Some types of repair kits include a heat source (usually a modified wood burning iron) and others tell you how to apply the required heat with a household iron. These repair kits also come with graining papers so you can match the texture of the original pad.

If you have a rather deep crack in your padded dash, you can use expanding urethane foam to fill the void before you apply the dash repair kit. This foam can be obtained through restoration supply firms, but it can also be found at home improvement stores. It is the same material that home owners use to fill cracks around pipes and electrical outlets. The foam is sprayed into the void where it expands and hardens. The excess is trimmed and sanded. These repairs aren't as good as a new dashpad, but they can be a big improvement over the damaged pad.

Modern One-Piece Dash

The third type of padded dashboard is the modern one-piece padded dash. This is the type of dashboard where the entire surface is padded, not just the top and leading edge. These dashes have been around long enough that many are cracked or otherwise damaged. Since they take up such a large portion of the interior, a damaged one can be a real eyesore. The fact that the padding goes everywhere means that damage can occur in very obvious places like around the radio or heating/cooling control panel.

It is possible to still get some of these entire dashboard assemblies from the new car parts department. There are some reproduction sources. Wrecking yards are a common source, but finding a perfect dashpad in a wrecking yard can be a real challenge.

The dashboard repair kits mentioned previously will work for minor repairs, but for a total restoration, you need to send the dashboard to a specialty dashboard repair company as described above. The workmanship is excellent and you end up with a dashboard that can't be distinguished from new. The downside is the expense. Depending on the size and condition of your dashpad, restoring it can cost several hundred dollars. Take this fact into consideration when you are tempted to buy a car with "only a little crack in the dashpad."

Dashboards

When looking for a good used dashpad, carefully check the underside as well. If a pad wasn't carefully removed, the mounting points can be damaged.

CLEANING GAUGES & TRIM

While you have your dashboard apart for painting or recovering the dashpad, you might as well clean everything associated with the dashboard. Dust and the effects of ultra-violet rays can make any dash look faded and old. The clear plastic covers for the gauges are often scratched from being wiped down with a dry cloth or paper towel. Most of these plastic components are easy to scratch.

Gentleness is the key to cleaning dashboard gauges and other components. You don't want to use abrasive cleaners. Harsh household cleaners can smear the white letters and numbers on gauges. Use the mildest possible household cleaners first. If you still have stubborn dirt, move up a step in the cleaning agents, but do so carefully.

Dust can often be blown away. A gentle form of air, such as a camera lens brush with the little air squeeze bulb will work well. The cans of aerosol air that are used on cameras and computers will also supply air that is more gentle than that from a compressor. Some parts of the dash can handle compressed air, but you need to be especially careful around the gauges. Use common sense first and force as a last resort.

You need to be careful about which components you get wet. Items that can be submersed can be cleaned in a gentle solution of mild soap and water. Household glass cleaners work fine for gentle cleaning, too.

Chrome Trim

Besides the gauges, you also need to be careful around the plastic chrome parts. This "chrome" is applied in very thin coats so it can be rubbed off quite easily. If your plastic chrome parts are in very poor condition, you can have them replated. Look in hobby publications such as *Hemmings Motor News* for companies that specialize in this process. If you have a popular vehicle, you can probably purchase reproduction plastic chrome dash bezels and other trim items.

True metal chrome items or stainless steel trim pieces can be cleaned with standard metal polishes. The soft pink polishes such as Simichrome work well. Chrome metal

Sometimes you can find new old-stock dashpads at swap meets or through old car want ad publications like *Hemmings Motor News*. These unused pads can be rather expensive, especially if it is a pad that isn't being reproduced. It is common to find black pads, but they can be dyed to other colors if needed.

Dashboards

The windshield was removed in order to install a new headliner in this early Mustang, so installing a new dashpad was very easy. This type of pad has a big opening in the center for the metal speaker grille. That means there isn't much support so the area can crack if it isn't handled carefully.

are applying the marker. Some people use model paint to touch up needles, but if you get too much paint on the needle it can affect the readings. You also need to be very careful not to bend or break the needles when working on them.

The raised letters on many dashboards have either silver or white markings. They can be restored by carefully applying the correct color model paint. Some people use a toothpick to apply the paint, but I prefer a very small paint brush (a 001 brush will do nicely). The idea is to put a minimal amount of paint on the brush so there aren't any runs. A steady hand is a big help.

Electrical Connectors

When you reassemble the dashboard components, use extra care on the connectors. Both the electrical connectors and the studs on the plastic bezels are fragile. Breaking these parts can slow down your project and

parts will polish up as long as they aren't too pitted. Minor pits can be removed by gently rubbing them with fine bronze wool. Stainless steel is the easiest trim to restore. Even badly dulled examples will polish up like new.

Plastic Lenses

Dull gauge lenses can be greatly improved with a plastic cleaner and polish. I have had success with Meguiar's No. 17 Plastic Cleaner followed by Meguiar's No. 10 Plastic Polish. Clean the lens first with a mild detergent so that you aren't rubbing unnecessary dirt into the lens.

Gauge Faces

The needles and writing on gauges often fade with time and exposure to the elements. If you are careful, you can restore these items. Whenever possible, use a marking pen to touch up gauge needles. You can find a wide array of marker colors at art supply stores. Day Glow Red or Orange marking pens work great for most gauges. Put a piece of paper between the gauge and the needle when you

If there is stainless trim around the edge of the dashpad, it can usually be restored with some metal polish. If these parts are missing, the reproduction industry makes many of these parts. The plastic gauge cover can be cleaned with plastic polish.

DASHBOARDS

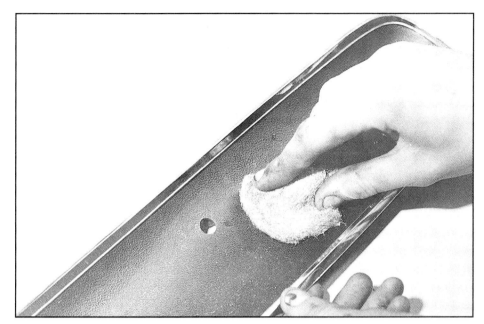

Some dashboard components (like this Mustang glove box door) can be moderately scratched or faded. In those cases, a light coat of touch-up paint can make the part like new. There was a texture to this glove box door, so the old paint was lightly scuffed with bronze wool to avoid damaging the texture.

generally ruin your day. The plastic studs usually won't tolerate very many applications of the fasteners. If you break off a plastic stud, you can make a replacement with a similar size metal screw. Cut the head off the screw and use epoxy glue to secure it to the bezel base.

GLOVE BOXES

A final touch for a dashboard restoration is the glove box. Many glove box liners are made of cardboard. They can easily deteriorate under too much junk being crammed into the glove box for too many years. Many restoration parts companies offer replacement glove box liners. Removing and replacing these liners is a fairly straightforward process.

In many cases you can also get new locks, new hinges, and even reproduction glove box doors. These improvements don't cost very much or take very long to accomplish. Even if your present latches are in good condition, check to see that the latch components are properly aligned. A little adjustment work will eliminate having to slam the glove box door. If you have a non-locking glove box, now would be a good time to add a lock.

Many owners of older cars and trucks want to install modern sound systems without cutting up the original dashboard. The glove box is a convenient location for extra stereo equipment. The original cardboard liner may not be up to holding a heavy equalizer or cassette/CD deck. The solution is to build a sheetmetal liner.

If you want to build the same size liner as the original, you can disassemble the cardboard liner and use it as a pattern for the new metal liner. If there is room behind the existing glove box, you could expand the size of the box. Even if you can't do the metal fabricating work or welding, you can save money by making the patterns. You might want to include some ventilation holes if stereo gear is what you intend to place in the new glove box. Consult a stereo store for recommendations on air flow requirements.

If your vehicle has a center console, the same restoration techniques that apply to the dashboard and glove box apply here. Center consoles often have hard to reach fasteners and interlocking parts so take your time when disassembling them. Refer to your interior shop manual for the location of fasteners. ■

The outer edge of the glove box door was shiny so it was taped off. A simple spray can was used to apply a light coat of new paint. Some dashboard components may be semi-gloss so check before you buy the paint.

Seat Upholstery Kits 8

Seats are the main element of any interior restoration project. Installing a seat upholstery kit can save you considerable money over the cost of having a professional upholstery shop do the work. The excellent quality of pre-sewn kits makes it possible for you to do professional quality work at home. Even the first time enthusiast can install seat covers with a minimal number of specialized tools (and the tools are very affordable). There are tricks and techniques to make the job easier and the results better. I will cover those items in this chapter.

Most upholstery seat kits are so well made that even the professionals use them. If you took your car to a shop, chances are great that they would use the very same kit that you can get from a mail-order restoration company. The biggest advantage that shops have over you is that they can custom make covers in special fabrics that aren't available in kit form. If you wanted to over-restore your car, a shop could replicate the factory vinyl covers in expensive leather. Without the ability to sew, you are limited to the vinyl reproduction seat cover kits.

Since the sewing is already done, installing seat covers is basically a

When duct tape starts to be the dominant upholstery material, it is time to install new seat covers. These Mustang seats look terrible, but a set of reproduction seat covers will make them look like new.

Seat Upholstery Kits

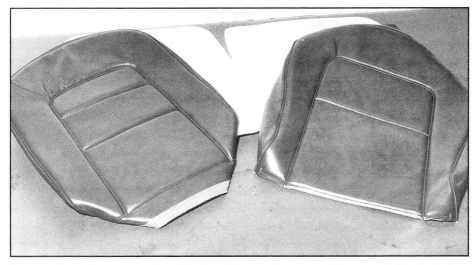

Here are the major components of the Corvette seat; the top cushion cover (left), the bottom cushion cover (right) and two new foam seat buns. The seat covers were sewn exactly like the factory original covers using identical material.

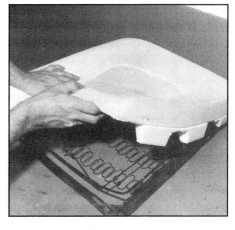

Minor seat cushion damage can be repaired, but the easiest way to regain the seat's original form and comfort is with new foam seat buns. The reproduction seat buns are shaped just like the factory ones.

remove and replace operation. If any of the underlying structures are worn or damaged, they need to be fixed before the covers can be installed. Most of these structural repairs are relatively easy to do.

Usually the biggest difference between a professional seat cover kit installation and work done at home is the quality of the fit. Professionals know all the tricks for making seams line up and avoiding wrinkles. You can overcome most of these obstacles by spending a little extra time and effort.

REPRODUCTION SEAT COVERS

There is a wide variety of excellent quality reproduction seat cover kits available. The vast majority of these kits are exact reproductions of the factory colors, materials, and styles, but there are some custom kits available as well. If you have a newer vehicle you can check with the parts department of your local new car dealership to see if factory replacement covers are available.

The quality of the current reproduction seat cover kits is generally quite high. I suggest dealing with an established parts company. You can talk to owners of similar cars to find out which upholstery kit they used and how they liked it. If you live in a large enough city that has specialty restoration supply businesses, check out the upholstery kits there. I like to get seat cover kits at a local outlet whenever possible so I can inspect the covers before I take them home. If there are any quality problems (or incorrect parts) it is easy to return the covers to the local business.

I have also had good luck with mail-order upholstery kits when there wasn't a local outlet. I try to determine which company actually makes the kits (as opposed to those companies that just retail them) and buy from them whenever possible. In cases where the kits are made by wholesale-only companies, I try to stick with well established companies with excellent reputations for customer service. Ask around at local car shows about which mail-order companies are the best. Other restorers are always eager to tell you who the good companies are and who to avoid.

When you order a reproduction seat cover kit, the company will most likely want to know the exact year, model, and trim number of your car. Simply stating that the car has black bucket seats may not get the desired seat cover kit. You might have a car that has previously had interior work or the body color was changed and the interior was done to match the new color.

The way to find out the factory interior trim level and color is to decode the information on the vehicle trim plate. A comprehensive factory shop manual will show the location of the trim plate or decal (common locations are the firewall, inner fender panel, driver's door jamb, or inside the glove box door). The manual should also tell you how to decode the trim data. Even if you don't know how to decode the information, the upholstery supply source will be able to understand the code and provide you with the correct seat cover kit.

When you get your upholstery kit,

Seat Upholstery Kits

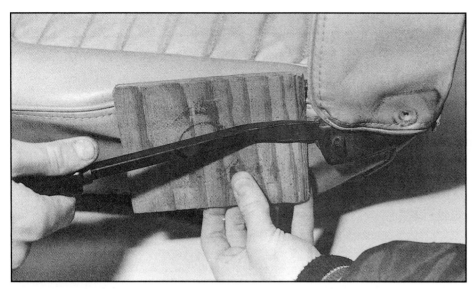

The seats need to be completely disassembled in order to install the new seat covers. Most upper seat cushions are mated to the lower cushion via a steel arm that goes over a pivot pin. Once any retaining clips or cotter pins have been removed, the upper arm needs to be pried off the pin. Place a piece of wood between the seat and the pry bar for better leverage.

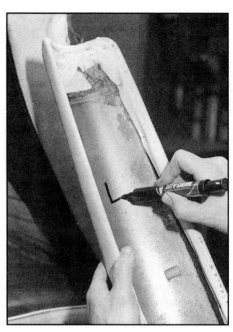

Many seats have end caps on the front seats. The caps are usually held in place with a couple trim screws. It is a good idea to mark the pieces (on the inside) so there isn't any confusion later as to where a particular part belongs.

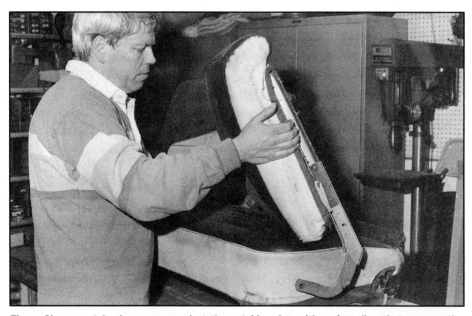

These Chevy seat backs are secured at the outside edge with spring clips that go over the seat pivot pin. There are plastic washers on each side of the seat back arm. The seat pivot arm needs to be pulled off the pivot pin. Then the whole upper cushion is rotated to free the inner pivot pin from the center mounting point.

check it carefully to be sure that you have all the required pieces and that you have the right kit. Mistakes can happen in the shipping department. Many vehicles looked similar for several continuous years, but each year was almost always a little different. The seat base could be the same for years, but subtle trim, stitching, and padding changes distinguish the different year covers. For example, a '66 Mustang seat cover will fit on a '65 seat base. The look is similar, but not correct. I have seen seat kits where the seat fronts were correct but the back seat panel was for the next year. That is why you should use your reference materials to double check that you have the right parts. Once you start punching holes for the hog rings, you won't be able to return the kit.

DISASSEMBLY

The old seats need to be removed from the vehicle in order to do a proper job of installing the new seat covers. Most seats are secured to the floorpan with a small number of bolts or clamps. By sliding the seat all the way forward and all the way back, the front seat fasteners should be obvious. If the floorpan has threaded holes you won't need to gain access to the fasteners underneath the vehicle. Fasteners that are reached from the bottom of the floorpan usually have plastic or rubber plugs to keep

Seat Upholstery Kits

moisture and debris out. If the vehicle was undercoated, you may have to locate the plugs and chip away the undercoating in order to remove the hole covers.

Rear seats are usually secured with a combination of clips and sheet metal screws. The screws often have big flat washers to distribute the load. Rear seat clips usually engage some of the bottom support rods. When you try to lift out the seat cushion, you should be able to feel where the bind is. Experiment with pushing the seat cushion back and up to clear the clip. The upper rear cushions are often fastened with screws at the bottom and clips up by the package tray. If you have any doubts about how to remove seat cushions, consult a factory shop manual.

Take care not to damage any other interior parts when you remove the seats. The front seat mounting bolts can snag and rip the carpet.

It is nice to disassemble the seats on a spacious workbench. I like to cover the work area with a big sheet of clean cardboard. I also use old (but clean) throw rugs to protect the upholstery while working on the seats. In the process of installing seat covers, the seats need to be moved around a lot to gain access to the various hog ring securing points.

The rear seat cushions don't need any disassembly (other than removing the old covers) unless you have a car with a fold down rear seat. If so, carefully mark and store all the folding hardware. Fold-down rear seats usually have stainless trim around the edges and carpeting on the back side of the seats. It is a good idea to draw a diagram of how the trim pieces fit the seat back.

FINDING SEATS

Usually a car will have the original seats, even though they may be in a poor state of repair. As long as you have the basic seats, restoring them with reproduction seat cover kits is no problem. The snag comes when the original seats are missing or severely damaged. Many popular collector cars have gone through cycles where they were modified. These modifications often included discarding the stock bench seat in favor of aftermarket bucket seats or buckets from some later model car. Incorrect bucket seats aren't just a problem with former bench seat cars. Non-original bucket seats often find their way into cars that were factory equipped with bucket seats.

Since cars with bucket seats are generally newer than bench seat cars, finding restorable seats isn't as difficult. You may still have to look a while to find a good seat at a reasonable price. Many cars use the same seat base for both the driver and the passenger's. For obvious reasons, the driver's seat is always much more worn than the passenger seat. If you can buy a passenger seat, the foam and inner springs will probably be in good condition. The seat tracks can be different for the two sides of the car so be sure to get the right sliding mechanisms and their related components.

Wrecking Yards

Wrecking yards can be a good source of affordable used seats; however, unless the wrecking yard is one of those highly organized ones that dismantles every car and stores the parts inside the warehouse, the seats can quickly get wet and damaged out in the yard. For this reason, you might ask the counterperson to give you a call when a possible donor car comes in. The sooner you get to wrecking yard interior parts, the better their condition will be.

The difficulty of finding decent seat bases for many collector cars and trucks means that you should be careful about discarding any seat parts. If you have a marginal seat base and you find a better one, don't be too quick to throw out your old seat. Another restorer may want your old seat because it is better than his. Also, some of the parts may be useful in saving another marginal seat.

Besides the difficulty of finding good seat bases, the various seat trim items can also be scarce. If the seats have wraparound bottom pieces or corner pieces for the seat back edges, these parts often get misplaced. Finding the upholstery to cover these trim pieces isn't any problem, but you need the metal base section. Some older bench seats used plastic wrap around pieces to cover the sliding mechanisms. These plastic parts are very prone to cracking and other damage. You may be lucky to find an unbroken set in the wrong color, but the plastic can be painted.

Seat Upholstery Kits

Old hog rings are often rusty and their sharp ends can damage tires if they are left on the garage floor. Keep a small container like a margarine tub nearby to hold the old hog rings. Pliers or diagonal cutters work well for removing the old hog rings. A twisting motion will quickly loosen the hog ring from the seat base.

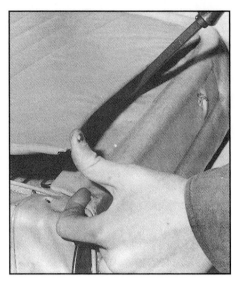

Some upper seat cushions have a separate rear panel that is secured with spring clips. A big screwdriver or pry bar can be used to pop the clips out of their mounting holes.

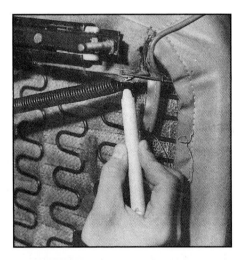

To ensure that the seat tracks work properly when the seat is reassembled, mark the location of the various springs, rods, levers, and fasteners.

Front Seats

The front bench seat or buckets need to have the sliding mechanisms and trim pieces removed. Take notes on how the various springs and sliding tracks are mounted to the bottom of the seat. You can use chalk to mark spring locations and which mounting holes are being used. It may seem simple when you are taking the seat apart, but it can be much more confusing when it comes time to reassemble the seat. If the seat tracks are rusty or in poor working condition, restore them before re-attaching them to the seats.

Separating the upper and lower front seat cushions usually involves prying off the pivot arm or hinge on the outside of the seat. These pivot arms are often hidden under chrome, plastic or upholstered trim panels. Some of these pivot arms can require a lot of muscle to get the arm over the pivot pin. Any pry bars should be backed by a block of wood to protect the upholstery. This isn't as critical during disassembly, but a misplaced pry tool can puncture the upholstery during re-assembly. When the outer hinge is clear of the pivot pin, the inner pivot virtually falls off.

Remove Seat Covers

You may decide to remove all the old seat covers at once, or take them off as you recover each cushion. The benefit of only removing one cover at a time is that you can use the old cushions for reference if there is any question about how a seam fits, etc. You could also continue to drive your car while the back seat and the passenger seat are recovered.

If this is the first time you have ever installed new seat covers, I suggest starting with the rear cushions. Their design is usually more basic than the front seats. If the installation is less than perfect, the rear seat isn't used as much as the front seat and mistakes won't be as obvious back there.

Remove Hog Rings

Turn the cushion upside down on the workbench. You will notice that a series of hog rings (loops of heavy wire) secure the upholstery to the seat framework. You can cut the hog rings with a pair of side cutters or diagonal pliers. You can also twist the hog rings open with a pair of needle-nose pliers or regular pliers. Try both

Seat Upholstery Kits

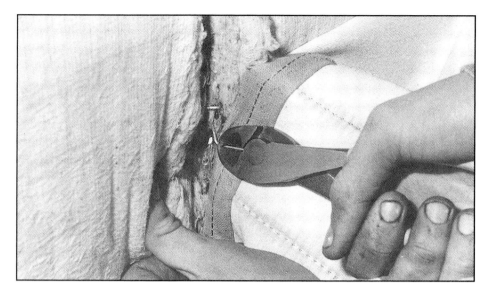

The outer hog rings are easy to spot, but there are also plenty of less obvious ones. Any place that the upholstery is recessed is a likely spot for a listing wire and hog rings. The backside of the front backrest where the cover is "rolled" and the "horseshoe" recess on bucket seats are common hog ring locations. If the seat cover won't come off easily, there are still some attached hog rings.

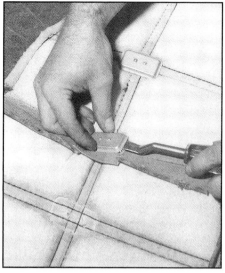

Seats that have any type of metal emblems or trim on the front side need some type of fastener on the back side. These press-on clips are the retainers for upper cushion trim buttons on a '70 Mustang seat cover. The clips and trim buttons need to be saved for use on the new seat covers.

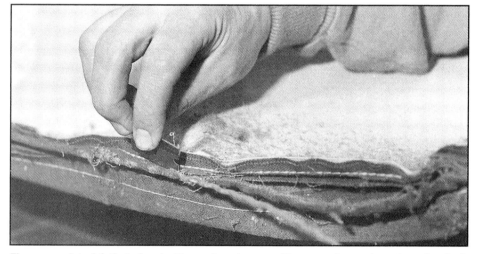

There are a lot of listing wires inside most seat covers. They are often rusty and can be stuck to the thin cloth listing sleeve. The quick way to remove rusty listing wires is by cutting the listing with a single edge razor blade. The listing wires need to be reused so don't lose any of them.

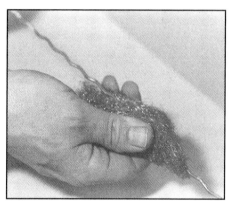

The old listing wires should be cleaned with steel wool, sandpaper, or a Scotchbrite pad. Clean wires are easier to install in the new seat covers. You don't want any rust on the new covers and listings.

techniques to see which one you prefer. You may find that each technique works well on different parts of the seat.

Besides the obvious outer edges of the seat cover, there are often hog rings in other locations. If there is any kind of recessed design to the upholstery, chances are excellent that a listing wire and hog rings were used to form the shape. A clue to where listing wires are located is on the new seat covers. Anywhere that you notice a sewn pocket for listing wires means that the original cover is secured in that area.

When you are removing the old hog rings, pay attention to which ones you remove. Hog rings can also be used to secure burlap over the seat springs or to fasten the seat foam to the springs. You only want to remove the hog rings that secure the actual upholstery. In some cases the same hog ring secures the upholstery and the foam or burlap. If this is the situation, remember to install the new hog ring through all of the elements.

After the old seat cover has been removed, put it aside in case you need

Seat Upholstery Kits

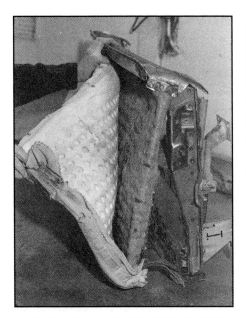

Under the front seat foam there is often a layer of burlap. The burlap protects the foam from the springs. If the burlap is frayed or ripped, replace it with a new piece.

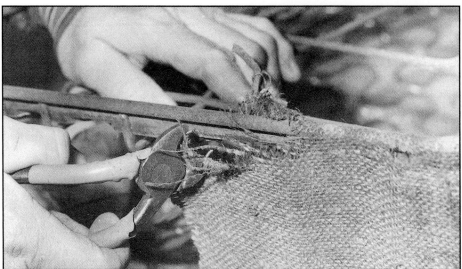

In some cases, like this corner section of burlap, the ripped fabric can be re-attached instead of replaced. Only a couple inches of the burlap deteriorated from moisture and rust, so hog rings were installed just beyond the rip.

it for reference (e.g., hog ring location and number of rings used). Remove any listing wires. If they are even the slightest bit rusty, clean them with steel wool, sandpaper, or a Scotchbrite pad. Rusty listing wires can snag and rip when installed in the new seat cover.

SEAT REPAIRS

After you remove the seat cover, inspect the seat base. You may find broken springs, deteriorated foam, loose burlap, or even broken seat frames. All of these elements need to be in good condition if you expect the new seat covers to fit right and look like new. Seat covers are similar to autobody painting. The exterior feature (seat cover or paint) is a reflection of the condition of the underlying elements. If the springs are broken or the foam is too compressed, the seat cover won't fit right. The new seat covers are sewn to fit correctly over seat bases that are in excellent condition. For example, if the seat foam is too worn out, the seat cover seams will end up in the wrong place. The seat cover can seem flat because the foam doesn't have its original shape.

Damaged or deteriorated foam is pretty obvious. When you remove the seat cover and it starts to snow foam flakes, you probably need new foam seat buns. If the old seat cover was ripped, that is a prime spot for foam damage, too.

Broken Springs

Broken seat springs can be a little more difficult to spot. If the seat has been sagging, chances are excellent that you have a broken spring. Springs can be broken down in the corners of the seat base. Carefully inspect the springs after you remove the seat cover. In severe cases, parts of the actual seat frame may be broken. A broken seat frame usually means you should find a better seat, but the metal framework can be welded back together.

Broken springs can be replaced. The more popular collectible cars have individual springs and entire spring assemblies available. There are also universal style springs available for general repairs. You may need to contact an upholstery shop to obtain repair springs. Springs can also be welded. A gas welder should be used for spring repairs because MIG and arc welders create too much heat for the relatively thin spring metal. The two broken spring ends need to be clamped together for welding. Be careful not to get any sparks on the burlap or other upholstery materials. A gas welder should also be used to repair any broken seat frames.

Burlap

Burlap or some similar material is often placed between the seat springs and the foam bun. If the burlap is ripped or otherwise damaged, it should be replaced. Check how the original burlap was attached to the frame or springs. Hog rings are the common method used to secure the burlap. Duplicate the original method

SEAT UPHOLSTERY KITS

Where burlap is more severely damaged, new burlap must be installed. Some seats use hog rings to secure the burlap, while other seats use trim adhesive to glue the burlap to the seat frame.

The edges of the burlap can be trimmed after it is glued to the metal frame. The burlap should be pushed into the recessed areas for a proper fit.

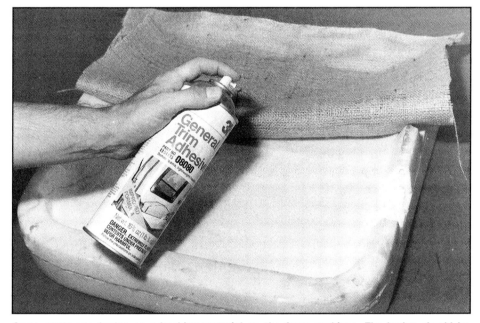

Some seats use burlap as a backing material on the foam cushions. The burlap should be attached with trim adhesive. A tip to make the job easier is to glue one half of the burlap at a time.

when you install any replacement burlap. Besides burlap, some seats use cotton muslin or similar fabric to protect the foam from the springs. If your seat cover kit doesn't come with this material, you should be able to get some at an upholstery shop.

Foam

Damaged or deteriorated foam is the most common seat base problem. The seat foam loses its original shape and size for a variety of reasons. The new seat covers are designed to fit the original foam dimensions. Foam that has shrunk or lost its original shape will make the new seat covers look loose and ill-fitting. Worn out foam means that the seats won't be as supportive and comfortable as they should be.

The first choice for fixing worn-out foam is a whole new foam seat bun. If new foam buns aren't available for your car, you can repair just the damaged sections. You may also be able to find a better foam base from a wrecking yard seat. For bucket seats, check out the passenger seats first since they receive much less use than the driver's seats.

Lower Bolster—The most commonly damaged section of any bucket seat is the lower bolster on the outside edge of the driver's seat. This area gets lots of abuse from getting in and out of the car. The outside part of the upper bolster also gets a lot of wear. The damaged section needs to be cut out. Upholstery shops have special electric foam cutting knives, but you can use an old electric carving knife, a serrated freezer knife (or any variety of similar serrated knives), or even a single-edge razor blade.

The foam should be cut out a little beyond the damaged area. If you can make the cuts square, it will be easier to cut and install the replacement piece of foam. Small holes (one inch

Seat Upholstery Kits

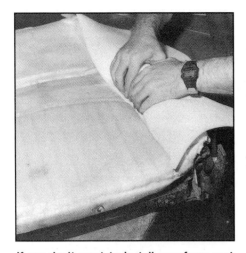

If you don't want to install new foam seat buns (or can't afford the added expense), compressed side bolsters can be built up by gluing a piece of quarter-inch or half-inch foam over the old bolster. Seat cushions that are too compressed will make the new seat covers sag.

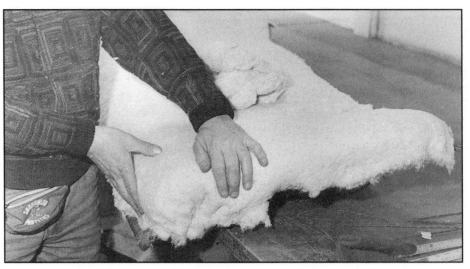

The cotton padding (on seats that use this type of padding) tends to get compressed or dislocated through years of use. If the seats are torn, the padding can disappear. Check the condition of the original padding. New cotton batting can be used to fill in any divots or low spots. Try to get it relatively smooth, but don't worry if it isn't perfect. The seat cover will help compress and smooth the cotton batting.

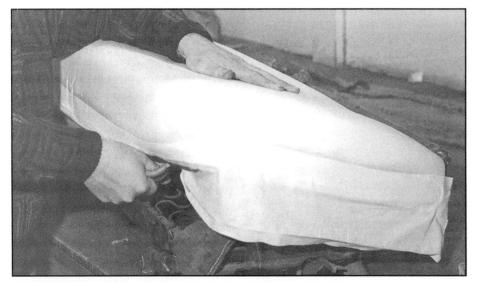

Some replacement foam seat cushions like this classic Chevy bench seat lower cushion from C.A.R.S. have a border of cloth (also known as a listing) attached to the foam. The cloth is pulled tightly around the springs and secured with hog rings.

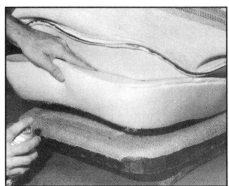

Some seat cushions are glued to the underlying burlap. If you do this, be sure that the positioning is correct before the glue sets.

or less) in the original foam can be filled with appropriate-sized plugs.

Foam Type—The replacement foam needs to be the same type as the original cushion. You can get this foam at upholstery supply shops and some general fabric stores. If you get the foam at a general fabric store, be sure that it is dense enough for auto upholstery. The foam at fabric stores is often lighter weight foam than what is used for automotive upholstery. If you only need a small piece of foam, a local automotive upholstery shop should be able to help you. They may even give you some of their scraps.

Attaching Foam—The foam patch pieces are secured to the cushion with automotive quality spray contact cement. Since the replacement foam is generally a little oversized, it will need to be trimmed to fit. Do the trimming after the patch has been glued to the base. The electric knives work best for trimming because they can gradually shave off the excess foam.

A trick to make the patched area as smooth as an unpatched cushion is to cover the whole bolster with a piece of half-inch foam. Besides making the whole bolster smooth, the half-inch foam will help compensate for any compression of the original foam.

SEAT UPHOLSTERY KITS

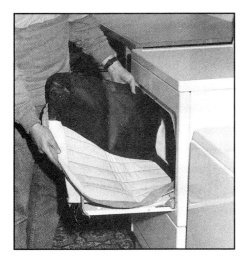

Reproduction seat covers come tightly packed in their shipping box which creates "box wrinkles." The covers should be unfolded and left flat in a warm room for a couple days to help relax the wrinkles. Another wrinkle removal trick is to put the covers in a clothes dryer for a short time. Monitor the covers to see that they don't get too hot.

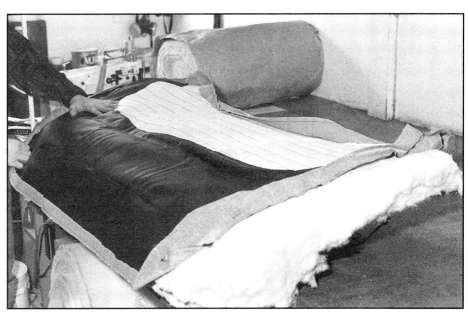

Install new seat covers for a trial fit to see if any adjustments are needed to the underlying cushion or padding. This cover was still a little loose, so another layer of cotton batting was added.

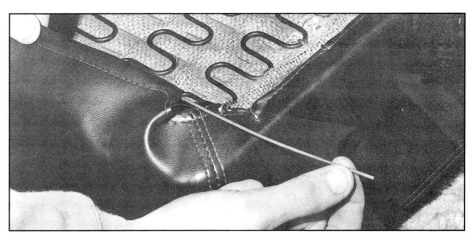

Before the seat covers are installed for the final time, remember to install all the listing wires in their proper locations. It is a good idea to wrap a little masking tape over the end of the listing wire. The tape will help prevent the wire from snagging inside the listing.

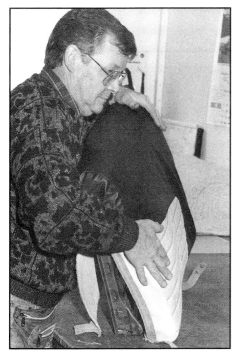

To get a good fit on the upper seat cushions, it helps to stand the cushion on the workbench and pull the cover into position. Use your hands to smooth the cover while working it down on the cushion.

Even if you only patched the outside bolster, both bolsters should be covered with the half-inch foam to keep the seat cushion uniform.

Even if you have seat cushions that don't need any repairs, you may want to consider applying a layer of half-inch foam anyway. The foam will help fill out the new seat covers for a nice, tight fit that makes the seat covers look like new.

When the seat bases are fully restored, you are ready to install the seat covers. Don't overlook the condition of the seat bases because a professional quality seat cover installation is dependent on the condition of the underlying seat base.

INSTALLING NEW SEAT COVERS

Installing the new seat covers is the reverse of removing the original covers. The listings (if used) go in the same locations, the cover seams line up in the same locations, and the hog rings go in the same locations. The

Seat Upholstery Kits

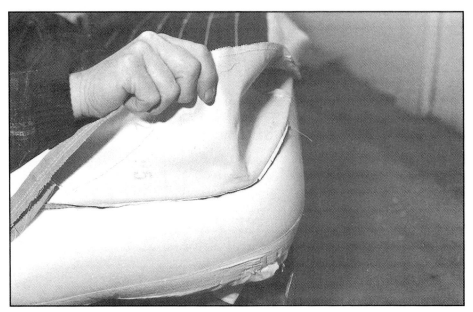

When you install new seat covers, start with the vinyl turned inside out. Line up the corners of the cover with the corners of the seat base. The small amount of seam material left over from the sewing should face down and rest on the horizontal part of the seat. This makes for a smoother look to the cover.

tricky part of all this easy sameness is getting a tight, wrinkle-free fit. A tight fit and the absence of wrinkles are usually the major difference between seat covers installed by a professional and an amateur.

There are several techniques that will help you achieve a good fit without wrinkles. The first step is to start with seat covers that are as wrinkle-free as possible. The seat covers usually come tightly packed in the shipping box. Take the covers out of the box several days before you plan to install them. Lay the covers out flat on a clean surface in a nice, warm room. If the weather is warm, you can lay the covers out in the sun (beware of dripping trees, birds, etc.). Use your hands to work out any wrinkles caused during shipping.

As discussed previously, a good seat cushion or base is essential for properly fitting seat covers. A base that is worn out or has deteriorated foam will yield poor fitting seat covers. For a good final fit, it is better to have a base that is a little big rather than too small. You can compress foam that is a little big, but shrunken foam will lead to wrinkles.

Professionals install seat covers by themselves, but an extra set of hands is a real asset for first time installers. The helper can stretch and hold the seat cover while you position and secure the hog rings. Your helper can also hold the vinyl tight while you use your hands to work out wrinkles.

Procedure

Start with the back seat cushion since it is usually less complicated than the front seat. It is a good idea to trial-fit the new cover. Check to see that the seams line up with the edges and corners of the seat. The seams should be a little inside their final position since you will be pulling the material tight. If it looks like the seams will end up lower than they should be, you may need to add some more padding to the seat base.

Tricks of the Trade—It is very normal for the new seat covers to be quite tight and even difficult to stretch over the base foam. Remember, tight is good; loose is bad. Some installers spray a light coat of silicone on the foam to make it slippery. Other people place plastic wrap such as Saran Wrap on the foam (or just over the corners). The plastic wrap makes it easier to pull the new cover over the foam. Some people use lightweight plastic

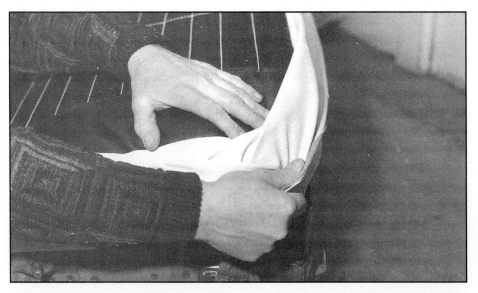

Then unfold the seat cover as you pull it down over the corners. Use one hand to pull and the other to keep the corner of the cover properly located on the corner of the seat base. The bottom edge of the cover needs to be pulled around the lower edge of the springs.

Seat Upholstery Kits

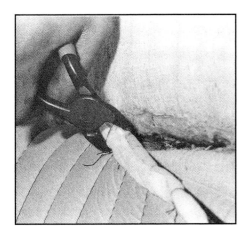

The new hog rings are installed in the reverse order of how they were removed. The inner listing wires are secured first. Sometimes it takes a lot of muscle to push the hog rings down through the openings in the seat base. If these inner hog rings aren't properly installed the fit of the cover will be off.

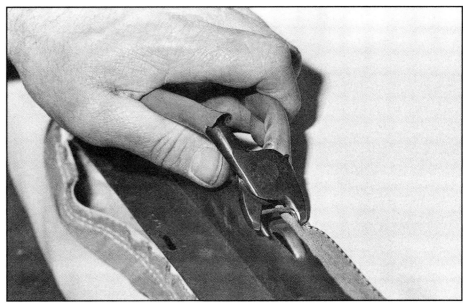

Some seat frames (or portions of the frame) have built-in loops for securing the hog rings. These loops make spacing easy. In areas where there aren't any loops, place the hog rings every 3 to 4 inches. If in doubt, consult the old seat cover.

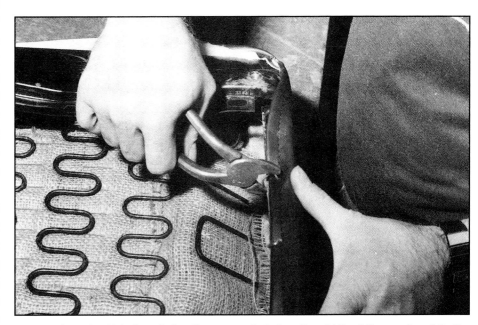

The hog rings should be installed at the corners first, then the middle of the panel, and finally, the areas in-between. Pull the covers as tight as possible without distorting the pattern of the seat covers. If there is excess material, fold it over and put the hog rings through the double layer of vinyl. This method makes the installation more secure and less likely to rip.

trash bags instead of plastic wrap. The plastic wrap that comes with dry cleaning can also be used. When the seat cover is in position, reach under the cover and pull out the plastic wrap or plastic bag. You can leave the plastic in place as long as it is thin plastic. Heavier plastics such as heavy-duty plastic trash bags could make an annoying sound if left inside the seat cover.

Another installation technique is to turn the new cover inside out. Then place it on the foam and roll the cover over the corners. When the cover is inside out you should notice a little extra material where the pieces of the cover were sewn together. This is called the *seam allowance*. On the lower cushions, be sure that the seam allowance points down as it goes around the corner of the cushion. This will help avoid wrinkles.

It is normal to have a few wrinkles in the seat cover when it is first installed on the base. Try pulling on the edges of the cover to see if the wrinkles disappear. The fewer wrinkles that there are in the early stages, the easier it will be to obtain a great fit later. There are ways to eliminate wrinkles even after the hog rings are in place, but try not to depend on these fixes.

Installing Hog Rings—When you are satisfied with the fit of the cover, it is time to start installing the hog rings. With the lower rear cushion, the usual starting place is the front corners, although some trimmers start in the middle and work their way out. You want the corners and seams to be in the right place and you don't want any

87

Seat Upholstery Kits

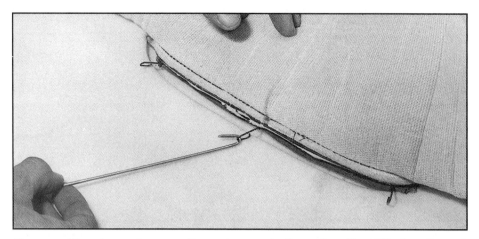

There are different ways to secure the seat covers to the seat cushion. This Corvette seat uses small metal hooks that are evenly spaced around the listing wire in the center "horseshoe" of the cover. A thicker gauge wire hook is used to grab the little hooks and feed them through the foam cushion.

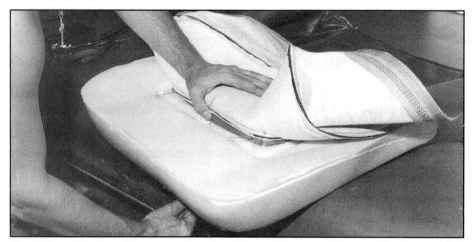

The seat cushion is positioned over the edge of the workbench and the small hooks are pulled down through the cushion.

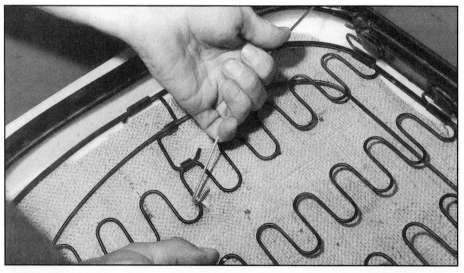

The end of the hook is placed around the nearest seat spring and the installation hook is released. Start at the top center of the "horseshoe" and alternate sides so that the seat cover is evenly positioned.

wrinkles. As you secure each area, use your hands to pull and smooth the fabric. Then install additional hog rings in a uniform manner.

Hog rings are generally spaced about 3 to 4 inches apart. If you are unsure about the spacing, check the hog ring holes in the old cover. Whenever possible, try to fold the material over so that the hog rings go through two layers of material. This makes the cover more secure with less chance of tearing the material.

When the front edge of the cover is basically secured, pull the material toward the rear of the cushion. Then install hog rings along the rear of the cushion. When you are satisfied with the front to rear fit, start work on securing the sides of the cover.

At any time during the hog ring installation process you can remove and reinstall rings if you aren't happy with the fit. Hog rings are cheap and easy to remove. Don't be alarmed if you reach down inside the foam to secure a hog ring and find that you missed the attachment point. This is a very common problem with listing wires that form the recessed areas of bucket seats. If you miss the attachment point, remove the dud hog ring before you try again.

Listing Wires—When listing wires are involved, start with the horizontal ones and then move to the vertical ones. Start at the center of the listing wire and work your way out to the edges. Since listing wires are usually in a rather visible part of the seat, take care when installing the hog rings. You want to grab the listing wire and the fabric sleeve that contains it, but you do not want to puncture the outer vinyl or fabric.

Pivot Points—The front seat covers usually go over the upper cushion

SEAT UPHOLSTERY KITS

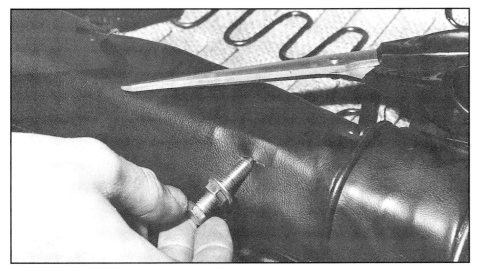

New seat covers don't come with any holes in them. You need to cut any trim mounting holes, rubber seat stops, or pivot bolt holes. Use an awl to locate the mounting hole and cut the smallest possible opening.

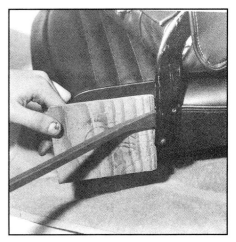

The use of a protective piece of wood is very important when it comes time to join the upper cushions to the lower ones. A misplaced pry bar could put a nasty rip in the new upholstery.

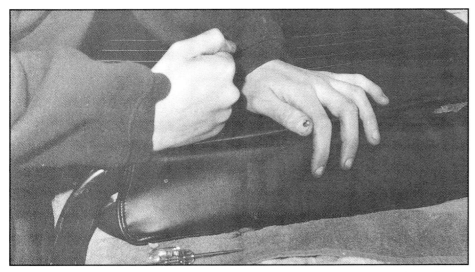

Seats that have a separate back panel use tension clips to secure the panel to the main part of the seat. Make sure that the clips line up with the holes and then tap them in place with your fist.

pivot points. You need to cut the cover to fit over the pivot points. Great care is required for this step. You need to be sure that the cover is properly positioned on the seat base before you cut the hole for the pivot. Cut the smallest possible hole. Most pivot points are partially covered by the upper cushion pivot arm, but too big of an opening can affect the tightness of the cover.

Attaching Upper & Lower Cushions—Another item that requires great care is the joining of the upper and lower cushions. The upper cushion pivot arm usually needs to be spread in order to fit over the pivot pin. Be extremely cautious when using any pry tools. You don't want to slip and put a hole in the new seat cover. A clean block of wood can be used for a leverage point, instead of prying against the upholstery.

Removing Wrinkles—You may want to deal with any wrinkles before or after you join the seat backs to the lower cushions. Sometimes minor wrinkles will relax with a little time, but to be sure, you should try to remove them at the time the seats are assembled. You can use your fingertips and the palms of your hands to work wrinkles to the edge of the cushion.

Heat will remove many wrinkles, but be sure to go slowly when using heat. A heat gun or an old hair dryer will both work. The hair dryer is slower than the heat gun, but the lower heat is better for beginners. As you warm the vinyl, use your hand to smooth out the wrinkle. If the vinyl is too hot to touch, you are using too much heat. Too much heat can permanently damage the vinyl.

Professionals like to use steam for removing wrinkles. They have special steam tools, but you can approximate the steam process with a water spray bottle. Lightly mist water on the wrinkle and use the hair dryer to dry the moisture and shrink the vinyl. It is much better to repeat any wrinkle

Seat Upholstery Kits

Some seat covers use these small clips to secure the lower edge of the bottom cushion to the seat frame. One loop of the clip grabs the seat cover seam and the other loop snaps on to the seat frame. Here is a lower seat edge clip being snapped on the seat frame of a Corvette seat. Space the clips evenly to avoid wrinkles in the seat cover.

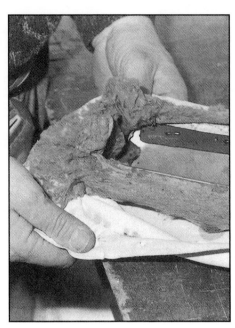

There is often cotton padding between the metal seatcap base and the vinyl covering. Remove the old covering slowly so that the padding stays with the base. You should be able to reuse the padding.

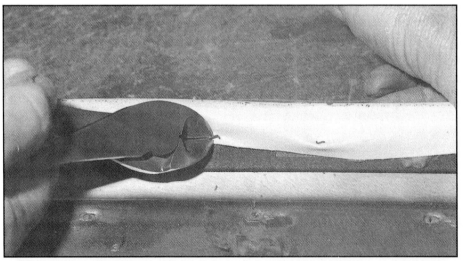

There are tacking strips on the back side of most seatcaps. Use side cutters to remove the old staples. Exercise care so that the tacking strip isn't damaged or disturbed.

removal process rather than use too much heat the first time. If you have any doubts about using a heat gun, find the most out of the way wrinkle and try to remove it before tackling the more noticeable wrinkles.

When you are satisfied with the fit of the seat covers all that is left is the installation of the various trim pieces. Some bucket seats have upholstered boards that cover the back of the seat. The clips that secure these panels can get damaged. Either get new clips or use a pair of pliers to straighten the old clips. Some seats use plastic trim pieces. These parts may need to be dyed or repainted to make them look new. Any chrome or stainless steel trim pieces can usually be greatly improved with a little metal polish.

Hopefully you put chalk marks (or made a good drawing) on the seat tracks so you can properly install the various springs. Put some white grease on the tracks so that they slide easily. Remember to feed any tricky seat belts into position as you secure the seats. The plastic center belt holders on bench seats can be especially difficult to get in place after the seat is bolted down.

The installation of new seat covers should greatly improve the looks of your interior. The process isn't difficult. If you pay attention to the little fitting details, you can achieve results that match those of any professional upholstery shop. ■

Seat Upholstery Kits

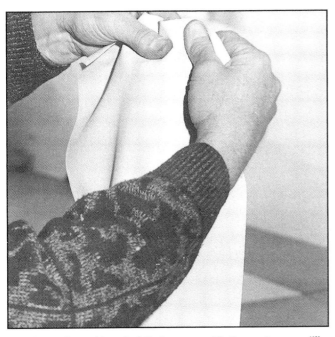

The piece of matching vinyl that comes with the seat covers (like on this classic Chevy kit from C.A.R.S.) needs to be cut to match the shapes of the old seatcap's vinyl. Secure one end of the new vinyl and pull the material taut to the other end. Stretch the material the long ways first. If there is too much excess material, trim it.

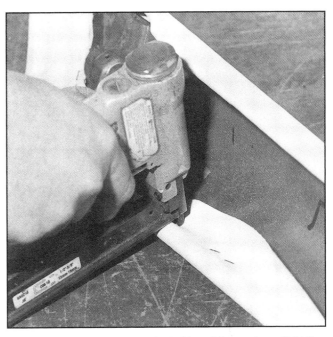

Then wrap the vinyl over the short sides of the seatcap. Hold the vinyl taut and staple it to the tacking strips. Be sure that the staples aren't any longer than original ones; otherwise they could penetrate the outer side of the shell.

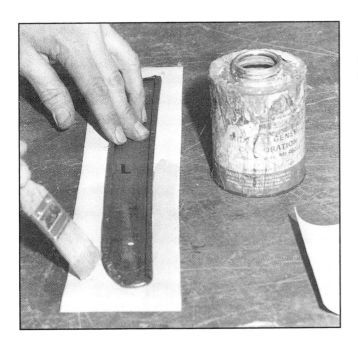

In this example, the two smaller trim pieces go on the inner side of the front seat backrests. The new vinyl is secured with contact cement rather than staples. Fold the excess vinyl over like you are wrapping a package.

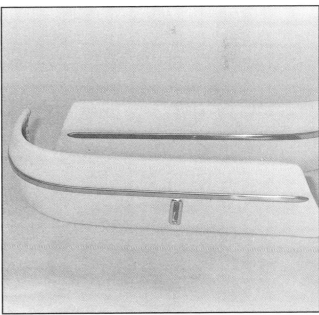

All new seatcap stainless steel trim pieces and seat hold-down brackets for classic Chevys are available from C.A.R.S. These recovered lower seatcaps look like they are brand-new.

Seat Upholstery Kits

After the new seat cushions are installed, let them be for a couple days. If there are any remaining wrinkles (especially with the heavier grades of vinyl) that don't relax by themselves, use a heat gun to warm and remove the wrinkles. Go slowly and carefully so you don't get the material too hot and damage it.

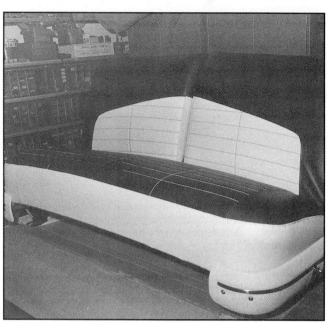

Here is the restored '56 Chevy bench seat sitting on the work bench awaiting installation in the car. The seat is a world away from the ripped, dirty seat that I started with.

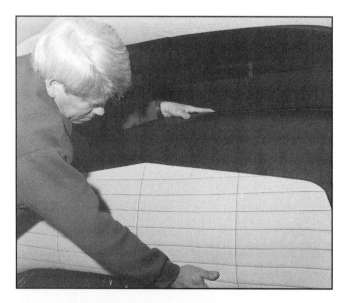

The seats are installed starting with the rear upper cushion. There are hooks along the edge of the package tray. There are four lower locating tabs. There are two recesses in the floorpan that hold two wire loops on the bottom seat cushion. Most rear seats have similar setups.

Installing new seat covers makes a tremendous improvement in an interior. With the great pre-sewn upholstery kits available, it is a project that can easily be accomplished in a home shop.

Kick Panels & Package Trays 9

Kick panels and package trays are seemingly small items at opposite ends of your interior that often get ignored. Restoring kick panels and package trays is usually quick and easy, so include them in your restoration plans. Since kick panels and package trays are flat, they are excellent places to master basic upholstery techniques.

With most popular cars and trucks, restoring kick panels and package trays is more of a remove-and-replace process than anything else. The reproduction items simply replace the old, worn-out pieces. You can use the reproduction panels as a base for making more deluxe panels if you would like.

KICK PANELS

As the name suggests, kick panels take a lot of abuse from shoes. The bottom edges can become water damaged on kick panels made of panel board. The panels can soak up water from the carpet. Once they get wet, it is only a matter of time before the deterioration process starts. The durability problems with panel board kick panels was eliminated when manufacturers switched to molded plastic kick panels. However, these plastic kick panels still will fade from sun exposure.

If reproduction kick panels are available for your car or truck, replacement should be very simple.

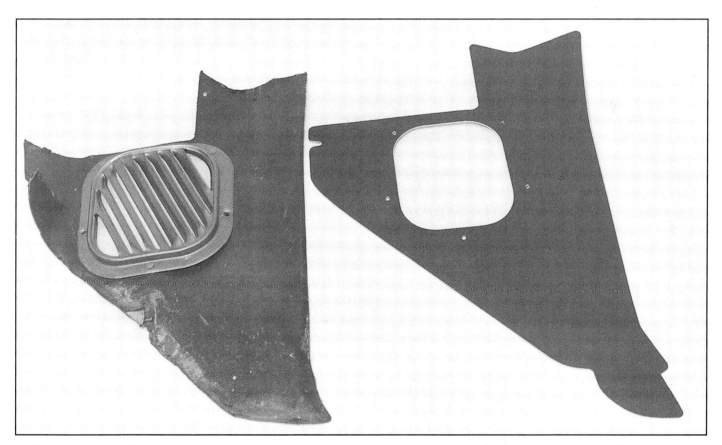

As the name implies, kick panels get kicked a lot. They are not the most visible part of an interior, but they are necessary for a finished look. The kick panel on the left is an original '56 Chevy unit. The reproduction panel on the right doesn't come with the air vent grille, so the original must be reused. Since the grilles are metal, they are usually only in need of a little cleanup and paint.

Kick Panels & Package Trays

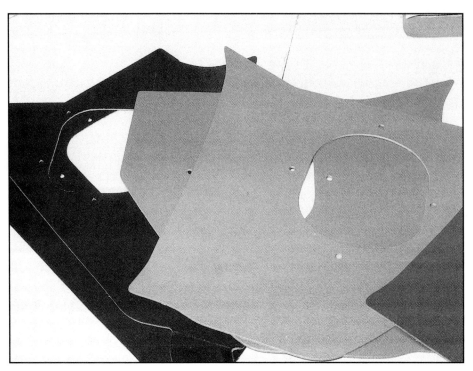

Many kick panels are not much more than colored panel board. They are available in a variety of colors for popular cars. They are one of the least expensive items you can get for your interior.

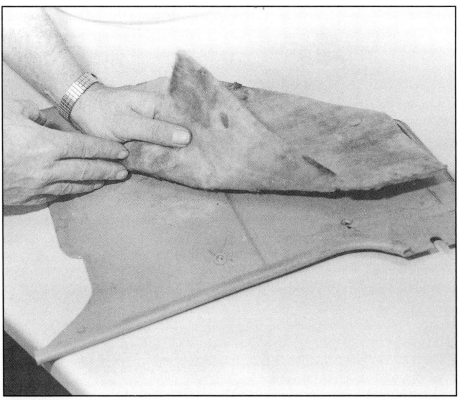

Some kick panels are molded plastic. These panels are sturdier than the panel board variety, but they can also get scratched and marred. Reproduction plastic kick panels are available, but they don't usually include the insulation material. Here the insulation is being carefully removed from the original kick panel.

There are usually just a couple screws or clips holding the panels to the body. Some kick panels are held in place by press-on windlace at the door edge. You may need to deal with air vents or vent control knobs. Sometimes radio speakers are mounted in the kick panels. Kick panels can hide wiring looms as well, so be careful to put the wires back where they belong and don't nick the wiring insulation.

The older style panel board kick panels are usually secured by a few screws too, but the vent grilles may be fastened with rivets. The replacement panel board kick panels are most often just the black board with holes for the vent grilles. You will need to reuse the original grilles. If your grilles are rusty, clean and repaint them. Install the grilles in the new panels. It might be difficult to find the exact rivets, but you could use standard rivets. You could also use rounded, carriage bolt style fasteners with the nuts on the inside of the panel. This way the exposed part of the bolt would be smooth, similar to a rivet. Unless you are worried about judging points at a concours event, you could screw the vent grille to the panel.

Restoring Plastic Kick Panels

Plastic kick panels that aren't in terrible shape can be cleaned and resprayed. Use a standard household cleaner and a scrub brush to remove all the old dirt and as many scuff marks as possible. Old toothbrushes work well for getting the dirt out of seams and crevices. When the panels are dry, treat them with a plastic prep solution. There are special solutions available at automotive paint supply

Kick Panels & Package Trays

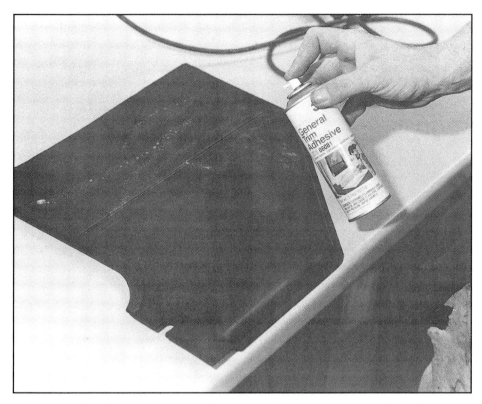

In this case, the whole interior color scheme was being changed, so new black kick panels were obtained. The back side of the panel was sprayed with trim adhesive.

components. Many people also cover their kick panels with carpeting that matches the floor carpet.

Covering Kick Panels—To cover the kick panels, use the panel as a pattern. Cut the material at least one inch bigger all around. You will want to fold the excess material around the edges of the kick panel for a finished look. Contact cement should be used to glue the vinyl to the kick panel.

If you use carpet, it can be too thick to fold around the edges. The best look with carpeted kick panels is to have a vinyl binding sewn to the edges. You could cut and fit the carpet sections and then have an upholstery shop sew the binding. You can glue the carpet to the kick panel or use stainless steel trim screws with matching washers.

If your vehicle doesn't have any existing kick panels, they can be made out of Masonite (available at lumber stores) or waterproof panel board. The Masonite is more rigid than the panel board, which makes it a good choice

stores for prepping plastic panels before dying them. Tell the counterperson what you plan to paint and they will lead you to the right products. The prep stage is important if you want the dye to stick to the plastic. Use the same color dye as the original panels (unless you are changing the color of the entire interior) and apply it in several light coats. The application of several light color coats will help you avoid runs.

Custom Kick Panels

Upgrading or modifying kick panels is relatively easy. You can add insulation behind the panels by either gluing the insulation to the panel's back side or just stuff in place as you reinstall the kick panels.

Adding upholstery to kick panels can be done without any sewing. A common technique would be to cover a panel board (or plastic) kick panel with vinyl that matches the door panels or other major interior

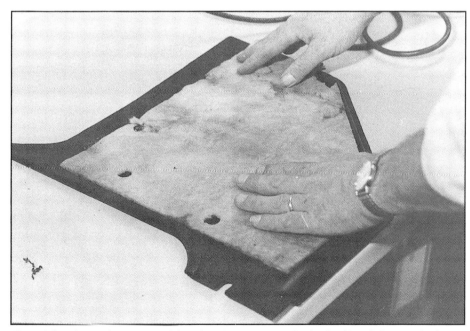

The original insulation was positioned and pressed on the back side of the panel. If your insulation is missing or too deteriorated to reuse, you can get new insulation material and cut it to fit.

Kick Panels & Package Trays

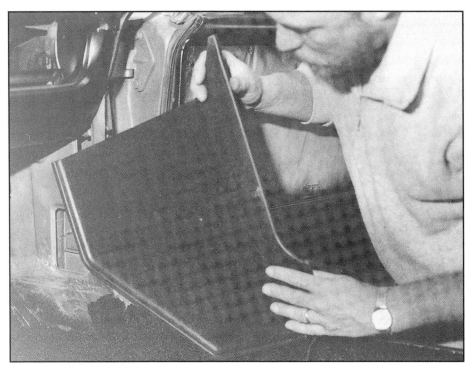

Kick panels are either secured with trim screws or snapped into position or a combination of the two. There can be metal locating tabs on the floor and cowl area. When installing the panel board style kick panel, be careful not to bend and crease the thin panel board.

There is often some type of windlace around the door openings. The windlace can be under or over the kick panel, so now is a good time to make any windlace improvements or repairs. Some vehicles use a metal retaining strip for the windlace.

This end view of ready-made cloth windlace shows how it is constructed. A piece of round foam rubber has a slit in the middle where a strip of heavy Kraft paper is inserted. The cloth fabric is wrapped around the foam and sewn to the paper. The flat edge is usually stapled to the tacking strips around the door.

for carpeted kick panels.

When covering the kick panels with vinyl or other fabrics such as tweed, the panels will look more professional if you include some foam padding. Either 1/4-inch or 1/2-inch high density foam will work. Glue the foam to the kick panel and then stretch the material over the foam. To make the corners smooth, notch the fabric on the back side of the panel. Just relieve the material enough to make the corners smooth. Don't cut past the back edge of the panel.

Speakers—If there is room between the kick panel and the body, you might want to install some stereo speakers. There are special thin speakers designed for kick panels. The speaker grille can be left on the outside of the panel or you can hide it.

To hide a speaker in the kick panels, you need allow air holes for the sound to escape. Cut the proper size opening in the kick panel to mount the speaker. If the panel has some foam padding, cut a hole in it, too. Tweed fabric is loose enough to let sound through, but vinyl should have some perforations cut in it. You could cover the whole kick panel in perforated headliner vinyl so you wouldn't have to worry about making the holes. You could also cover just the speaker with the headliner material and the rest of the kick panel with solid vinyl.

Storage Pouches—Kick panels also provide a handy location for small storage pouches or bins. There are a couple ways you can install storage pouches with little (if any) sewing. You might be able to turn up some suitably sized plastic storage bins in a wrecking yard. Look at small imported cars. They tend to have lots of neat little storage bins. Plastic bins can be cut to fit and glued to your

KICK PANELS & PACKAGE TRAYS

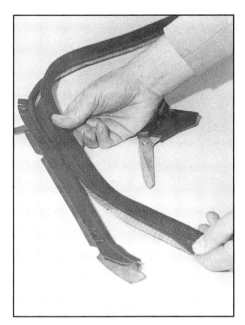

On cars with the metal retaining sections, the windlace needs to be inserted in the narrow channel. Sometimes this is a press fit, while other times glue is needed.

existing plastic kick panels or newly formed Masonite kick panels. Even if the gluing isn't a work of art, you can hide it with a vinyl covering.

To make vinyl storage pouches you need to make a pattern for the finished size of the pouch. Then make another pattern that is approximately two inches bigger all around. Cut out the vinyl to match the bigger pattern. Fold the vinyl to match the smaller pattern. Cut a hole in the kick panel vinyl that corresponds to the small pattern. Make sure that the hole is where you want the pouch to be. Mark the outline of the small pattern on the kick panel base.

To make an elastic opening for the pouch it helps to sew in a section of elastic. If you can sew this straight line on a household sewing machine, you could put the elastic inside the folded over excess vinyl and glue the flap to the inside of the pouch vinyl.

The folded-under excess vinyl from the large pattern can be glued to the kick panel (once it is in the proper location). Then if everything was lined up correctly, the main vinyl can be glued to the kick panel. You might want to place a couple rivets or staples inside the upper corners of the pouch to help secure it to the kick panel. This style of storage pouch won't be as sturdy as a sewn pouch, but it will still be fine for light-duty use.

PACKAGE TRAYS

Package trays are the shelves that are located above and behind the rear seat, between the C-pillars and under the rear window. They are more visible than kick panels so it makes sense to restore them. Package tray restoration can be as simple as slipping a new one on the shelf area. The ease of your package tray restoration depends on the type of unit used in your car.

Check your shop manual for specific details on how your package tray is secured. Some package trays fit under the sail panels that cover the inside of the C-pillars or the back of the roof. There can also be window molding trim that holds the package tray in place. The leading edge of the package tray can be secured underneath the back of the rear seat. In these cases, you will need to remove the rear seat cushion. Sometimes the rear side (or quarter) panels are also involved.

Package trays can be glued to the rear sheet metal shelf. If you have a replacement panel, don't worry about damaging the old package tray. If the old tray needs to be retained for a pattern, use a putty knife to separate the tray from the shelf.

Reproduction package trays are very simple to install. If you want to repaint or add upholstery to the package tray, the procedures are the

The package tray area of rear windows is subject to lots of heat. The package trays are easily faded or cracked. If the original package tray was glued in place, use a putty knife to remove the chunks of old tray and glue.

Kick Panels & Package Trays

You can simply install a new package tray, but with a little extra effort the underlying metal can be restored. Sand any rust and cover the metal with a rust preventative paint like POR-15.

same as outlined in the kick panel section. Many package trays had some type of insulation (often jute) under them originally. Whether or not your car came with any insulation, you may want to add some to help block any noise coming from the trunk. Besides jute or foam, consider some of the new high tech closed cell bonded foam (with the aluminized facing) insulation or the heavy composite rubber or plastic material. These materials can be found through street rod companies and stereo installation centers. Many of these new insulation materials have adhesive backing or you can use construction adhesive in a caulking gun.

Custom Package Trays

Most of the techniques discussed with kick panels apply to building custom package trays. If your vehicle doesn't have an existing package tray, you will need to make a pattern. Depending on the slope of your rear window, it can be a little difficult to get in there and measure correctly.

Make a Pattern—To make a pattern, first measure the width of the package tray at its widest point. Get a roll of Kraft paper or parcel wrapping paper. Lay the cut piece of paper on the rear shelf and mark the contours of the package tray. It is better to cut the pattern a little oversize. Lay the rough cut pattern on the shelf and make the necessary adjustments. When you are satisfied with the paper pattern, transfer it to something stiffer like poster board or cardboard. Check the fit of the cardboard pattern. It needs to be exact if you want the finished product to fit right.

Transfer the cardboard pattern to waterproof panel board or thin Masonite. Heavy duty scissors can be used to cut the panel board or you can use a utility knife. If you use a knife, place the panel board on a big, flat surface with some cardboard under it for the knife to cut in to. An electric saber saw with a fine tooth blade can be used to cut Masonite. You could also use a small coping saw.

Once you have a suitable base, you can add foam padding and top it with vinyl or other fabric. Package trays are a very common speaker location. You can mount speaker grilles on top of the package tray or hide the speakers as mentioned in the section on custom kick panels.

If you want to add a custom touch to your package tray, you can make a sculptured tweed panel as described in the door panel chapter. Use thin layers of foam to build up the designs and cover the whole panel with the tweed fabric. These tweed package trays work well for hiding stereo speakers.

Paying attention to the kick panels and package tray will give your interior a detailed, finished look with a minimal amount of time and effort. ■

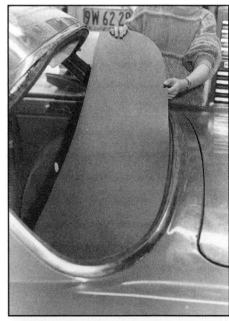

Package trays are usually plain panel board or panel board with a vinyl covering. Even with the rear window in place, installing a package tray is usually pretty simple. If speakers are part of the package tray, they usually need to be secured from inside the trunk.

TRUNKS 10

Although some may not consider the trunk a part of the interior, I believe otherwise. Too often trunks become a giant junk depository, which detracts from the car overall. A nicely restored trunk is less likely to be used as a rolling trash bin.

It is a rare car that still has it original trunk mat in mint condition, but reproduction companies make new mats for most popular cars. You can also buy bulk trunk mat material from mail order companies such as J.C. Whitney and cut it to fit your application.

Many vehicles use a trunk partition board between the back of the rear seat and the trunk. These panel boards are the same material as package tray panels and kick panels. Reproduction

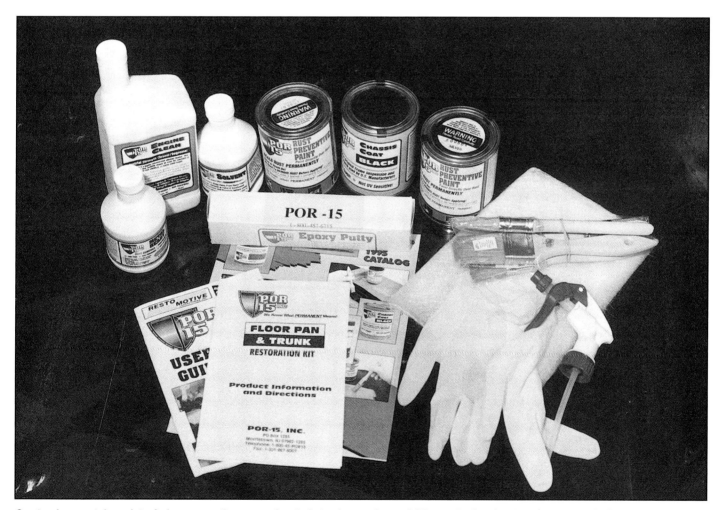
Car trunks can take a lot of abuse over the years. Granted, trunks aren't as visible as the interior, but they are part of a thorough interior restoration. If your trunk doesn't have large rust holes, the small holes, pinholes, surface rust, and pitting can be fixed with a POR-15 Floorpan and Trunk Restoration Kit. The kit contains all the materials, chemicals, paint and applicators needed for the job.

99

TRUNKS

After the trunk floor has been prepped and any holes filled, the whole area is painted with black POR-15 rust preventative paint. It can be sprayed on, but since the paint flows out nicely, a brush does a fine job.

parts companies stock them.

CLEANING & REPAIRING

The first order of business for any trunk restoration project is a thorough cleaning. Remove all of the stuff (or junk) that has accumulated through the years, and use a high powered shop vacuum to pick up the dirt. If your trunk mats are in decent condition, you may be able to wash them and recondition them with a vinyl conditioner. Remove the mats from the car. Use an upholstery cleaner (and a heavy-duty household cleaner on the greasy spots) and a scrub brush. After they are dry, use ArmorAll, STP Son of A Gun, or Meguiar's Intensive Protectant to help restore the original look.

Minor Rust Repair

While the mats are out of the trunk, check the condition of the trunk floor and the spare tire well (if so equipped). Rust is a big enemy of trunks. As unpleasant as it is to deal with rust, ignoring it won't make it go away. Rust will only get worse with time, so repair it as soon as it is discovered.

There are patch panels available for many cars. Unless you are an experienced welder, you will probably require the services of a body shop. If your trunk just has a few small holes or a lot of surface rust, you can fix it yourself. A rust repair kit like the POR-15 kit (used on the floorboards in the carpeting chapter) will fix small rusted-out areas and pin holes. The POR-15 will also prevent future rust problems because the paint is a rust inhibitor.

There are a couple major sources of rust inside your trunk. You should check them out to avoid future problems. Poor fitting, deteriorated weatherstripping won't properly seal the trunk. Excess moisture can also enter the trunk from the rear window area. Many cars have rust problems around the lower part of the rear window. Moisture can attack the package tray area as well as go straight into the trunk. Cars with vinyl tops are particularly vulnerable to this problem. Even if the vinyl top still seems serviceable, water can seep through the pores of the material. Use a strong light to inspect the underside of the area between the trunk lid and the rear window for signs of rust.

Rust can also enter the trunk from the bottom of the car. This is a particular problem in sections of the country with harsh winters, lots of snow and ice, and where salt is used on the roads. You can also add water to the trunk inadvertently by making it a habit to throw wet snow chains in the trunk. Keep the chains in a strong plastic bag.

Spatter Paint—If you are going to cover the trunk floor with a trunk mat, you may want to limit the amount of work you put into detailing the floor.

TRUNKS

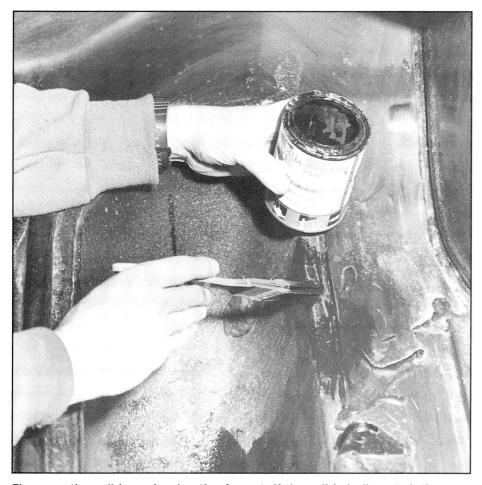

The spare tire well is a prime location for rust. If the well is badly rusted, there are replacement panels available but they require welding. This spare tire well was in good condition so it only needed an application of rust preventative paint.

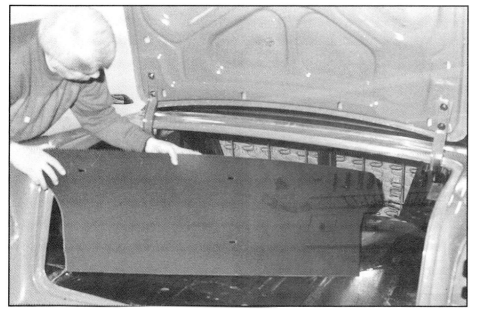

Reproduction companies offer replacement trunk mats and trunk boards. The trunk board separates the trunk from the rear seat. These boards are reasonably priced. This Chevy trunk board was secured by metal tangs on the back of the seat braces.

If you are a perfectionist, you may as well strip off the old paint, repair any rust problems, and apply new paint. Some trunks were painted in a spatter-type paint. You should be able to obtain this paint at your local paint supplier or hardware store. The Eastwood Company offers spatter paint in aerosol cans. After it is applied, the paint should be sealed with an acrylic sealer for best results.

Gas Tank

The top of the gas tank also serves as the bottom of the trunk in some cars. If you have rust problems here, you may need a new gas tank. If you just want to clean and recoat the inside and outside of your gas tank, the Eastwood Company offers products and kits for this task. If your trunk has a visible gas filler neck and hose, check the condition of these items and replace as required. For a proper restoration, be sure to use the correct style clamps. Many original hose clamps were the wire style, not the more commonly available worm-drive aftermarket hose clamps.

Seams

While you are cleaning and repairing the trunk, check all of the seams. If the factory seam sealer is missing or damaged, recoat it with fresh seam sealer. You can get either brush-on sealer or the type that is applied with a caulking gun. There are different colors of seam sealer, so if you want your car to be concours correct, do your research and get the correct product for the job. Show judges can also dock you points for using the wrong color weatherstrip adhesive.

TRUNKS

Here is the '56 Chevy trunk after cleaning, patching, and painting. There are reproduction rubber trunk mats available for these cars or you can install some carpeting.

Many trunks were originally coated with splatter paint. The Eastwood Company offers two-tone trunk paint in several color combinations if your trunk needs to be touched up.

Decals

Trunks usually had a couple decals or glue-on instruction notices about the proper way to use the jack. These decals are often quite shabby, but you can get reproduction ones for popular collectible cars.

Wiring & Taillights

While you are cleaning and detailing your trunk, check to see that all the taillight housings and wiring is in good condition. Now is a good time to check the fit and operation of the trunk latch and locking mechanism. The taillight panel can rust out on many cars. Debris such as tree needles, leaves, and dirt can collect around the taillights and between the bumper and the rear valance. This debris retains moisture which leads to rust. The taillight panel might look okay from the outside, but when you remove the taillights, there can be a lot of rust around the openings.

CARPETS & UPHOLSTERY

You can go beyond the simple replacement of factory-style trunk mats and add carpeting or upholstered panels to your trunk. Since most stock trunks are so bare, it doesn't take much work to upgrade your trunk. Many newer model cars have carpeted or upholstered trunks. These trunks are quieter and offer a more inviting cargo area. It also helps to have a more luxurious trunk since many newer cars have fold-down rear seats for increased carrying capacity.

To carpet your trunk, you will need to make patterns and cut the carpeting

TRUNKS

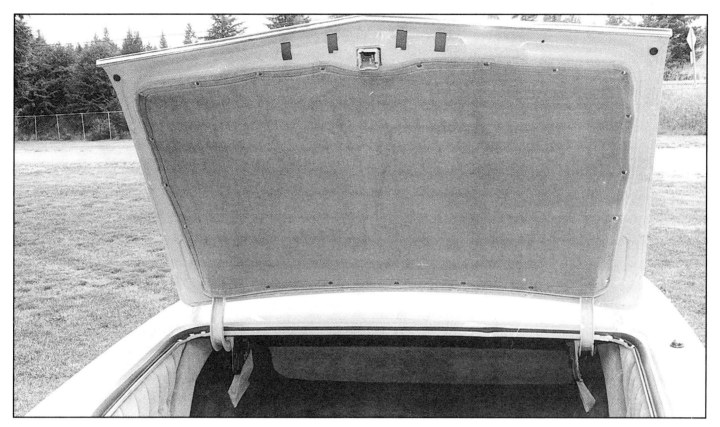

This trunk lid has a custom made carpet liner. You can make similar covers by gluing carpet to the trunk lid. Some people glue small sections of carpet inside the recessed areas of the lid instead of using one big cover.

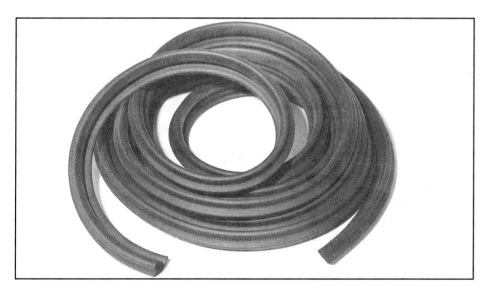

One reason that trunks develop rust problems (and that unpleasant moldy smell) is poor weatherstripping. Deteriorated weather-stripping allows water into the trunk. Reproduction weatherstripping is readily available and easy to install.

to fit. It is a good idea to cut the carpet slightly oversize first, and then trim it after you have placed it in the trunk. You can always make it smaller, but you can't go the other direction. Carpet looks better with bound edges, especially on low nap carpeting, which is often used in trunks. You can get an upholstery shop to sew on the binding after you cut and fit the carpet. Carpet with a deeper nap will hide the seams better than low nap carpet.

If you pick a carpet with a stiff backing, it can be difficult to fit into corners and curved areas. Unbacked carpeting is easier to work with. You can get unbacked carpet at an upholstery supply store or through an upholstery shop.

Installing Carpet

Carpet can be glued right to the metal surfaces of the trunk. Gluing will give a nice, tight look to the installation, but you may want to make the carpet removable for cleaning. A snap kit can be used to mount snaps in the corners of the carpet and in the corresponding

103

TRUNKS

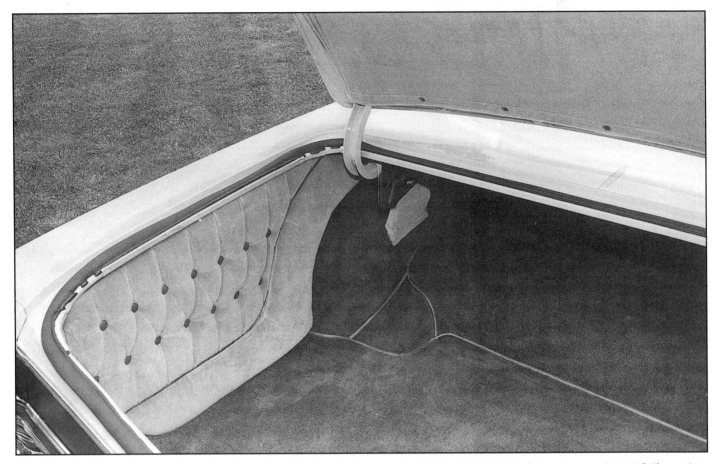

Custom trunk side panels can make a nice change from the plain, factory panels. This trunk has panels that were sewn to match the custom interior, but less elaborate panels can be made by gluing fabric or carpet to the stock panel board inserts. Very thin plywood veneer can also be used for the backing boards.

locations underneath. Or, you can use hook and loop fabric strips for easy in-and-out maintenance.

Upholstered Panels

To upholster the side panels of the trunk, you can cover the existing panels (if the vehicle is so equipped) or you may need to build your own custom panels. These panels can be made from panel board, Masonite or thin wood such as a veneer. The wood bases will be more durable than the panel board, but the panel board is easier to work into odd-shaped corners.

You can also make your custom-fitted panels different than the true shape of the trunk. You can avoid some of the odd shapes and make the panels straighter. These panels will be easier to upholster and if you make them removable, you can hide the usual trunk clutter behind the panels. To be really clever, you could install a couple hidden hinges for an access panel. Custom-built panels will usually need some type of bracing or support system. You may have to custom cut and shape sections of plywood so that it can be glued to the inside of the trunk. A light, easy to work with wood such as pine could be employed to make a frame like those used for kitchen cabinets.

Once you have the backing boards or panels fabricated, it is just a matter of covering them with the fabric of your choice. As with kick panels, a covering of thin but dense foam will give the panels a luxurious look. The sculptured tweed over foam technique described in the door panel chapter will also work well on trunk liner panels.

Covering the inside of the trunk lid is a little difficult. The problem is the odd shapes created by the support ribs and the lock mechanism. Flexible tweed material can be glued right to the lid. The low areas could be filled with foam. Cloth-backed foam can be cut to match the shape of the trunk lid. Vinyl or fabric can then be glued to the foam and the whole pad can be attached to the trunk lid with snaps, trim screws (use short enough screws so that you don't puncture the trunk lid), or hook and loop strips. ■

Steering Wheels

11

Steering wheels aren't generally associated with upholstery, but they are an integral part of any automotive interior. You might not use your back seat or glove box everyday, but you won't get very far if you don't use the steering wheel. With all that use, steering wheels can easily become worn and/or damaged. Most steering wheels are a plastic outer surface over a steel core. The damaging effects of heat and ultraviolet rays make steering wheels brittle over time. Small cracks lead to big cracks.

There are three basic options when it comes to restoring your steering wheel. You can replace the old wheel with a new one; restore the existing wheel to like-new condition; or replace the stock steering wheel with an aftermarket wheel. If your current wheel is in excellent condition, you can get by with a thorough cleaning.

Steering wheels need to be repaired out of the vehicle. The first step is to disconnect the battery so you can remove the horn button without annoying the whole neighborhood. A wheel puller is required to remove the steering wheel. These pullers aren't expensive, but you can also rent them.

Steering Wheels

Steering wheels can crack and get discolored over time. While it is possible to restore a worn steering wheel, finding a new old-stock wheel is less work. The only problem is that the swap meet (or specialized parts dealer) prices are far more than the cost of restoring your existing wheel. The POR company makes a steering wheel restoration kit.

It is a good idea to use a torque wrench when reinstalling a steering wheel. This Mustang wheel wasn't cracked, but it wasn't the right color so it was repainted. For steering wheel torque specifications, consult the factory shop manual.

STEERING WHEEL REPLACEMENT

The first order of business for any steering wheel replacement or repairs is to disconnect the battery. You will need a steering wheel puller. They aren't very expensive and can be found at local auto parts stores. You can also rent them at tool rental outlets. Most wheel pullers come with a variety of bolts to fit into the threads inside the steering wheel hub. Be sure that you pick the matching bolts. The bolts should go in by hand at first. If you try to force in the wrong bolts, you can damage the hub and have a very difficult time removing the steering wheel.

New, original steering wheels are very difficult to find for classic cars of the '50s, '60s and '70s, and if you do find one, expect to pay plenty. Reproduction steering wheels are virtually unheard of, although horn rings and center caps are available for many vehicles. Swap meets and specialty wrecking yards are a good source for used steering wheels. So if you want to keep your vehicle equipped with a stock steering wheel, your choices are limited to finding a good used wheel or repairing your existing steering wheel.

While standard plastic steering wheels can be difficult to locate, deluxe wood grain (either fake or real wood) steering wheels are even more difficult to find (at least at a reasonable price). If you have always wanted the deluxe steering wheel for your car, you may want to restore your standard wheel and wait patiently until you find a great deal on a wood steering wheel. If you find a worn wood steering wheel at a swap meet, consider buying it if the price is

Steering Wheels

An inexpensive alternative to custom steering wheels is to install a better wheel from a different model vehicle. Many of the factory deluxe steering wheels are very attractive and they are reasonably priced for all but the most exotic wheels (e.g. certain Corvette steering wheels and wood rimmed Mustang wheels). If you are installing a Chevy steering wheel in another Chevy product, there shouldn't be any compatibility problems.

fair. Even wood steering wheels can be repaired. Restoring wood steering wheels at home isn't as easy as it is to fix standard wheels, but there are craftsmen who specialize in restoring wood steering wheels.

When searching for a replacement steering wheel, don't overlook different colored wheels. If you find a nice steering wheel that is a different color than your interior, you can have the wheel painted the correct color. A factory original color that matches your car is always best, but you can find some good deals on odd-colored steering wheels.

Upgrading

If you are not shooting for a first in show at a concours event, where originality counts, you might want to consider upgrading to a steering wheel from a later-model car of the same marque. Same model (or same manufacturer) steering wheel swaps are very easy since you don't need hub adapters. Many Chevrolet owners trade their plain steering wheels for ones from Corvettes or Super Sport models. There are even deluxe truck steering wheels that will greatly improve the look of a base model truck.

These deluxe factory steering wheels can easily be found at wrecking yards and swap meets. Except for some high-demand Corvette steering wheels, most of these wheels are quite affordable. I have found some super deals on brand-new deluxe truck steering wheels at swap meets because van and truck conversion companies often install custom steering wheels. They sell the stock deluxe wheels for extremely reasonable prices. These conversion companies also have some great deals on stock seats that they replace during the customization process.

The only thing to be careful about with swap meet steering wheels is to be sure that the wheel will fit your car. A 50/50 swap meet guarantee usually means fifty feet or fifty seconds, whichever comes first.

When you are ready to remove the

This steering wheel swap is subtle, yet effective. A '59 Chevy Impala steering wheel was installed on the factory steering column of this mildly modified '54 Chevy Bel Air. Swaps like this are great and easy to accomplish.

Steering Wheels

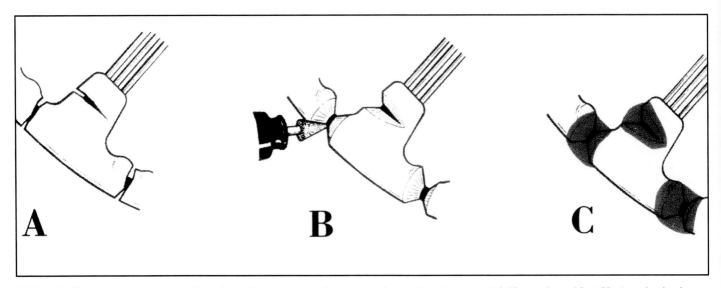

(A) Cracks like these are common with early model plastic steering wheels. Sun and heat caused shrinking and cracking. Most such wheels can be restored inexpensively without recasting them. (B) A Dremel tool or small files can be used to "hog out" cracks to accept plastic filler. (C) Mix up only enough epoxy filler to fill a couple of cracks at a time. Overfill cracks, then sand them to shape using 60-grit coarse sandpaper when they harden to the consistency of hard cheese. Finish with 360-grit sandpaper. Drawing by Jim Richardson.

original steering wheel, check to see that the vehicle's wheels are straight. You want the new steering wheel to be centered when you install it. An extra check is to mark the steering shaft spline at the twelve o'clock position with a marking pen or chalk.

STEERING WHEEL RESTORATION

Restoring an existing steering wheel is often your least expensive, and in many cases, only option. At first glance, the task may seem difficult, but it really isn't all that hard. Most steering wheels either develop cracks or the color fades. The cracks can sometimes expand enough to be called chips or gaps, but they can still be repaired. The only time color is a big problem is with steering wheels from the fifties and sixties that were either transparent or translucent. Paint repairs depend on opaque paint.

Repairing Cracks

There are several steering wheel restoration kits available from companies like The Eastwood Company and the POR-15 Company. The core of these kits is a two-part epoxy filler that is used to fill the voids in the steering wheel. People have been known to repair steering wheels with regular body filler, but catalyzed epoxy putty is preferred.

Enlarge Cracks—The first step is to identify all of the cracks of the wheel. Then you need to make them worse. This seems contrary to the idea of repair, but the epoxy needs a large enough surface to bond with. If the steering wheel has separated completely, then that is obviously a large enough gap, and you can skip to the prep and fill phase. However, most cracks need to be worked out a bit so the epoxy has an area to grip.

A variety of cutting or grinding tools can be used to enlarge the cracks. Most steering wheel repair kits suggest cutting a "V" notch in the wheel. This can be done with a die grinder, a hack saw blade, a file, a rasp, or even a drill bit. A small drill bit such as a 1/8-inch bit can be used sideways as a grinder. Even smaller drill bits can be used to drill a couple "tunnels" on the sides of big gaps. The filler will go into these areas to act as "anchors" for the main repair. Just be careful not to do more damage than good with these anchor holes.

Apply Epoxy—Mix the epoxy putty according to the directions and apply it to the gaps. It is fine (actually a good idea) to build the filler up above the surface of the steering wheel. The excess material will be removed during the shaping and sanding portion of the repair. Work the filler or epoxy into the gap. A popsicle stick will work well. You want the filler to be pushed all the way into the gap. You don't want any air pockets.

Sand Filler—After the filler has dried completely (in some cases this can be as long as two days), you can shape it. A combination of small files, rasps, and sandpaper will do the job. Go slowly so that you don't cut too deep and have to refill the area. You

STEERING WHEELS

Many custom steering wheels have padded center covers. They are usually a press fit. Since they serve as the horn activator, be sure the battery is disconnected.

CUSTOM STEERING WHEELS

A quick way to give your car or truck a more luxurious, sporty look and feel is with a custom steering wheel. Most of these aftermarket steering wheels are leather covered, but some are made of beautiful wood, such as mahogany. Custom wheels are almost always thicker and better padded than skinny, plastic stock steering wheels. Once you enjoy the comfort of a custom steering wheel, you will wonder how you ever got along without one.

Colors & Styles

Custom steering wheels come in a wide variety of styles and colors. Most are black, gray, or tan, but you can also get custom dyed leather wheels to match (or contrast) almost any interior. There are even custom steering wheels with billet aluminum centers to match custom billet wheels. These steering wheels are made by the same companies that produce the road wheels.

Custom steering wheels are typically secured to the hub with Allen head screws. The screws need to be tight, but be careful not to damage the fasteners. The wire and clip immediately to the left of the Allen wrench is the horn connection.

also want to avoid cutting into the good part of the steering wheel.

Use sandpaper in steps. Start with a coarse grit in the 80 to 150 range, then go to a medium grit in the 320 area, and finish up with a fine grit in the 400 to 600 range. Do the final sanding wet. If you are in doubt about what grit sandpaper to use, follow the directions with the steering wheel restoration kit. The idea is to use ever finer grits to remove the scratches left by the coarse grit sandpaper.

Prime & Paint—The wheel should be primed with a primer surfacer to help fill any tiny imperfections. After the primer has dried, wet sand the steering wheel with a fine grit sandpaper. Consult your local automotive paint distributor for the best paint for your wheel. You want a paint that won't harm the plastic. Depending on the gloss characteristics of the paint, you may need to buff the paint to achieve the desired new wheel appearance.

If you don't have the equipment for spray painting, many paint stores can make a custom aerosol spray can with the correct paint. You can also buy inexpensive aerosol units that have a glass jar attached. These spray units are great for applying small amounts of paint.

Simulated wood steering wheels can have a slight amount of "grain" in them. If you cover up the grain with repair filler, you can get the texture back by dragging a hacksaw blade along the plastic. A coarse brush should be used to apply the wood graining paint. You may need to do a little more scraping with the hacksaw blade if the paint obscures the grain.

Restoring your old steering wheel may be a little time consuming, but the material and equipment costs are negligible compared to the cost of a new one.

Steering Wheels

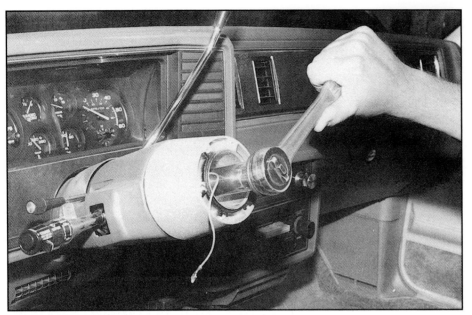

Most custom steering wheels require an adapter hub. These hubs are made for specific vehicles, but the actual steering wheels are the same to keep manufacturing costs down. Be sure that the vehicle's wheels and tires are absolutely straight before the hub adapter is installed.

Installation

Installation of custom steering wheels is pretty simple as long as you have the proper hub adapter. Most custom steering wheels have a universal hub pattern, but a model-specific hub adapter is required to go between the custom steering wheel and the end of your steering column. The hub adapter also houses the horn connectors.

When you are ready to remove the original steering wheel, check to see that the vehicle's wheels are straight. You want the new steering wheel to be centered when you install it. An extra check is to mark the steering shaft spline at the twelve o'clock position with a marking pen or chalk. When you remove the stock steering wheel, note where the little horn springs and wires belong. These connector pieces need to be retained and properly installed in the hub adapter in order for your horn to work. While you have the hub apart it's a good time to clean all the contacts.

When you install the custom steering wheel and hub adapter, be sure the wheel is properly positioned in the hub at twelve o'clock. Torque the hub retaining nut to the specifications in the wheel directions. Some hub retaining nuts also use a safety clip, so don't forget to install it. Even when you think that the hub is securely tightened, it doesn't hurt to re-check it after a few months of driving. I have seen installations where the hub retaining nut backed off from use.

STEERING COLUMNS

While you are restoring your dashboard and steering wheel, you might as well restore the steering column. If you have always wanted tilt steering, now would be a good time to install it. If the turn signals don't work properly, or the self-canceling feature doesn't work, you can get replacement parts and install them once the column is apart.

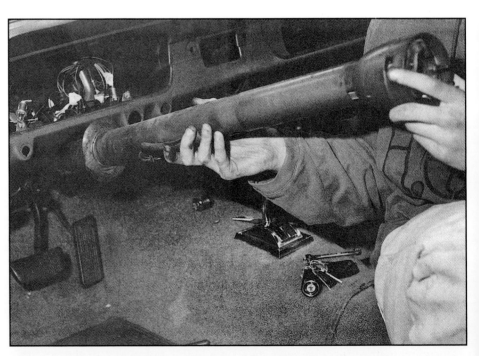

When a dashboard receives a color change, it is usually a good idea to paint the steering column to match. This is easier to do on older vehicles that have a removable outer tube. If the transmission gear selector is mounted on the column, take notes as to how it was positioned.

STEERING WHEELS

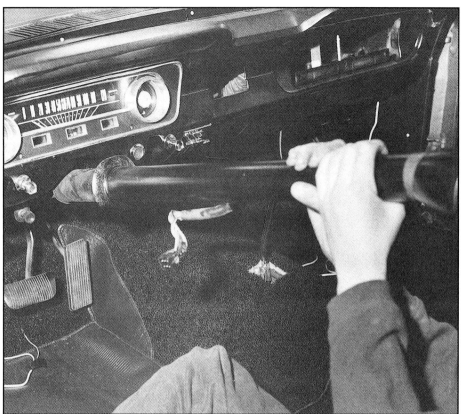

Installation of the freshly painted steering column is basically the reverse of the removal, except that you need to be more careful about column components staying in their correct location. Also, pay attention to any wires so that they don't get pinched.

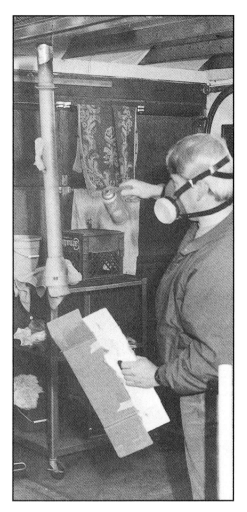

It is possible to paint steering columns while they are still inside the car, but a more thorough job can be done outside the vehicle. This way it is easier to do a superior cleaning and prep job, as well as getting good paint coverage in all the odd crevices.

Steering columns can have a lot of small parts and wires jammed into relatively tight quarters. Take notes of how things fit as you remove them. Be very careful not to drop the tiny screws down inside the column housing.

If you just want to repaint the steering column, you can do it with the column still in the vehicle. It will be easier to get a quality paint job if you remove the column from the car. Be sure that the steering and wheels are centered before you remove the column. Make reference marks so that everything will go back together correctly.

The procedure for painting the steering column is the same as for any other metal interior component (see Chapter 3). The factory service manual should have the correct paint number for your column. If you don't have the manual, your local paint store should be able to help you.

Installation

To install a tilt steering column, try to find a similar vehicle in a wrecking yard and get the unit out of it. Many of the steering columns produced for GM, for example, were made by the same manufacturer and share many components. Still, they can be different lengths so getting a column out of a similar car can avoid a lot of potential problems.

If you have an older car or truck that never came with tilt steering from the factory, you can adapt columns from never vehicles. There are even companies that make custom tilt columns. These columns can be had in bare steel or polished stainless steel. You may need to use special adapters to connect your steering box to the tilt column. The companies that sell the new columns can help you get all the right parts for a safe installation. Check the ads in custom truck and street rod magazines for sources of these new tilt steering columns. ∎

Custom Seats

12

Custom seats are one of the quickest and easiest ways to transform a plain interior into a deluxe driving environment. Once you have enjoyed the comfort and support of custom seats, you will hate to drive a vehicle with standard seats. Depending on the brand and model of the seat, you can get superior lateral, lumbar, and leg support. These factors make everyday driving much more pleasurable, and greatly improve comfort on long trips. If you like to drive aggressively through twisting mountain roads, the secure placement of your body will allow you to focus on driving instead of trying to stay in one place.

Many people balk at custom bucket seats because of their price. Yet, these same people don't mind spending the same amount of money on custom wheels. Granted, the wheels will impress more strangers, but why not treat yourself instead of other motorists? A good way to justify the cost of custom seats is that you can keep the seats for a long time. Most of these seats rely on adapters to mount them to specific vehicles. When you

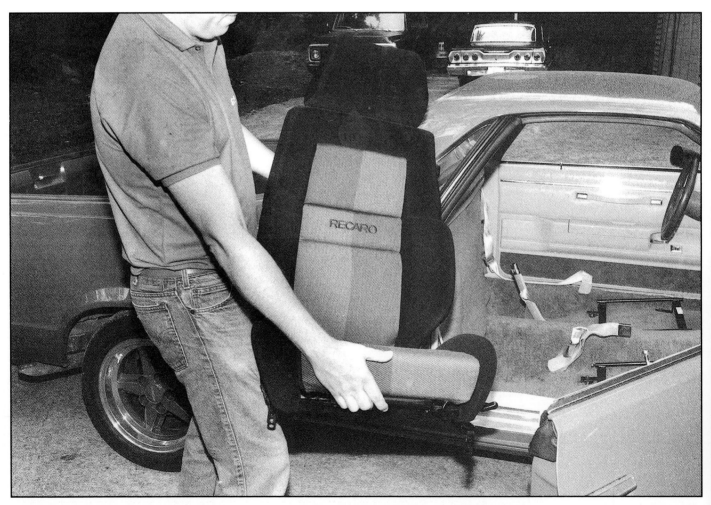
Custom seats are one of the quickest and easiest ways to make an ordinary interior as comfortable as an expensive sports car. High quality seats like these made by Recaro offer outstanding comfort and the bolsters help keep you from sliding around when cornering.

Custom Seats

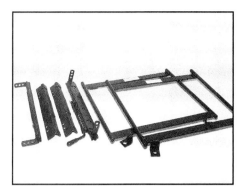

Most custom seats are a universal design. They can be fitted to a wide array of vehicles with specialized seat mounting brackets (right). Some custom seats can be mounted to the factory bucket seat hardware. The items on the left are parts of the seat adjustment sliders.

The custom seat mounting frames are designed to match up with the factory mounting points. The factory fasteners are used.

change vehicles, you can retain the custom seats and just get new mounting brackets.

There are many manufacturers of custom sport seats. The prices and quality of the different seats vary. You should do some investigating prior to making such an expensive purchase. Talk to owners of these seats. Check out cars and trucks similar to yours at car shows and see how the seats fit. These seats often sit closer to the floorpan than stock bench seats. There can be clearance problems with the transmission tunnel and the door sills.

Some custom seats can be ordered in a modular system. This way you can match the seat back and base that best suit your driving needs, budget, and available space. Seats with lower bolsters are easier to get in and out of. Generally these seats are less expensive than the seats with the bigger bolsters and more adjustable features. The seats with the high bolsters really cradle you, but they are more difficult to access. If you do a lot of errands and short trips; you may want the versions with easier accessibility.

The more deluxe seats have knobs for adjusting lumbar support and the tightness of the bolsters. The really high end seats have power adjustments. Some seats have little built-in air bladders that can be pumped up or deflated to adjust the lumbar support. You can even get some seats with optional heaters.

Custom seats are available in a wide array of colors and fabrics. You can probably find a combination that is close to your existing interior or a seat that goes well with what you have. Some companies will cover their seats with factory matched fabric at an additional cost. High end seats are

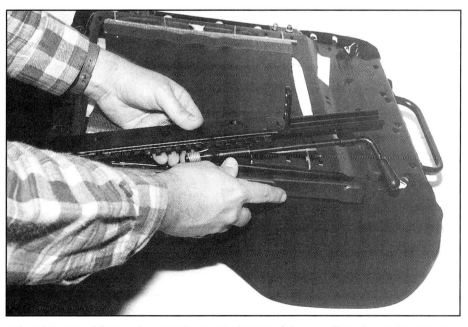

Adjustable seat sliding tracks are bolted to the bottom of the seat. Then the tracks are bolted to the mounting frames on the vehicle floorboards.

Custom Seats

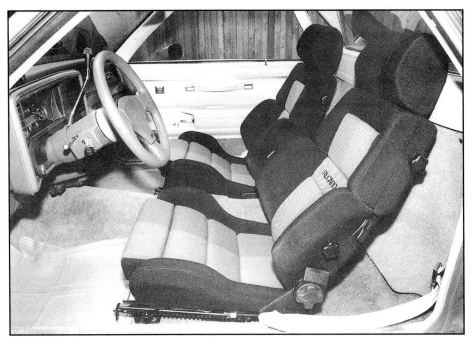

Here are a pair of Recaro sports seats installed in a late model El Camino. The knobs on the side of the seat are part of the multi-point adjusting system. The front section of the lower cushion is also adjustable for proper thigh support.

The more expensive custom seats have knobs for adjusting lumbar support and side bolster width, but lumbar support can be increased or moved around by inserting specially shaped pieces of high density foam.

Most custom seat manufacturers will sell extra material so that the back seat can be upholstered to match the custom front seats. This Mustang fastback has custom front sports seats with matching upholstery on the back seat. This makes the interior look like a factory installation. This is not a bolt-in operation. The rear seat needs to be recovered at an upholstery shop.

At the budget end of the custom seat spectrum, you can get fiberglass or molded high density polyethylene plastic racing bucket seats. These seats tend to be a little narrow, but their light weight makes them excellent for race cars. They usually come with thin vinyl seat covers.

available in leather. The leather seats are very luxurious, but they cost considerably more than cloth-covered seats. Solid colors are your best choices for a decent match and being able to use the seats in another vehicle. Black seats not only go well in many interiors, they are the easiest color to resell.

Another option for matching the new seats to the rest of your interior is to obtain a few extra yards of custom seat fabric. Many seat companies will sell you the matching fabric so you can have the rear seat reupholstered to match the bucket seats. I have had this done and the results were spectacular. You can also use the matching material to recover the door panels for a totally integrated look.

Custom Seats

A budget seat swap is to get the deluxe seats from the same vehicle or a similar vehicle from the same manufacturer. Deluxe velour covered split bench seats with a nice armrest like this example will bolt in a similar vehicle with the standard bench seat. Often something like a Buick seat will bolt into a Chevy that was built on the same corporate body platform. Seats that aren't a direct bolt-in can be used if the seat tracks are modified.

SEAT SWAPS

A popular alternative to custom aftermarket seats is to swap in seats from another make or model vehicle. Many of the factory seats on newer cars and trucks are almost as nice as custom seats. The seats from sports cars and luxury cars are the most popular for swapping. Some imported car seats are so popular that the cars get stolen just for the seats. Highly popular seats can command premium prices at wrecking yards, but there are many less popular but still excellent seats for reasonable prices.

When searching for seats don't just look at the sporty cars. Many sedans (imports, especially) have excellent seats. Even trucks have nice bucket seats, but they can be a little big for some car installations.

You may be able to find seats from a newer model of your car. Late-model Camaro or Mustang seats in an earlier car have the added appeal of keeping your car "all Chevy" or "all Ford." Seats that have an identifying logo can look out of place in a dissimilar brand vehicle.

If you have a base model of a particular car, you can often improve your interior by finding seats from the top-of-the-line version. If you are lucky enough to find the same color, you can save a lot of extra work. The best thing about upgrading your seats from a same make car is the ease of installing the new seats.

Modifications

When you swap seats from newer models into older cars and trucks, you may need to do some fabrication work in order to attach the new seats to your floorpan. Try to get all the mounting brackets that belong to the new seat. It helps to do some thorough measuring of your car before you go seat hunting. When you know the space limitations of your car, you will be able to make well-informed choices during your seat search.

Seat Base—The older the recipient vehicle and the newer the donor seats, the greater the chances that you will have to raise the new seats. Take measurements of the old seat height before you remove it. If you were less than thrilled with the stock location, you can alter the new seat location. Seat bases should be made of sturdy metal tubing. The use of old 2x4 wood blocks is unsafe and unacceptable. You might be able to bolt together a suitable base, but it is best to use one that is welded. If you don't have good welding skills, have the work done by a qualified weldor. You don't want the seat base to separate in an accident or sudden stop.

Even if you farm the welding out, you should still do the initial layout and design. You could make a mockup base out of wood and then have a metal fabrication shop copy the design in the proper gauge tubing. By making a wooden mockup, you can check to be sure that the seat will be at the right height.

Finding good used seats at a wrecking yard can be difficult unless the seats are stored inside. Seats left out in the wrecked cars can get greasy, ripped, and water damaged. Smart yard operators know which seats are desirable and usually keep them inside.

Custom Seats

High performance racing seat belts and shoulder harnesses are a nice addition to high performance cars. The seat belts use special mounting bolts that fit in the stock seat belt bolt holes. Racing style shoulder harnesses need to be securely mounted and at an angle that isn't harmful to the occupants. There are shoulder harnesses that are designed to mount to the roll bar. Floor mount shoulder harnesses can interfere with using the back seat.

Trucks

Trucks are great candidates for seat swaps. They are roomy and the stock bench seats often leave much to be desired. Besides bucket seat swaps, some late-model bench seats (which often have fold-down armrests) are perfect fits in earlier cabs. For example, owners of older Chevy pickups can use late-model deluxe Silverado bench seats. Many of these newer truck seats actually increase leg room since the new seats are thinner, but still offer great support. The newer seats have high quality foam that gives excellent support in a thinner seat.

A popular trend for customized trucks is to install the power bench seats of luxury cars such as Cadillacs, Lincolns, and Chryslers. If the luxury seat is the correct width, the seat and power base can be bolted into the truck.

Common Problems—There are a couple things to watch out for if you are contemplating such a seat swap. A potentially big problem (that too many builders tend to overlook) is the angle of the luxury seat's backrest. Remember that the luxury seats came out of cars with a back seat, so having the backrest angled into the rear passenger area wasn't any problem. Most stock truck seats are rather upright due to the space limitations imposed by the rear of the cab. People who try to install a luxury seat with a greatly reclined backrest angle will find the lower part of the seat uncomfortably close to the pedals and the dashboard.

Another potential problem with swapping luxury bench seats into trucks is the width of the donor seat. The measurements of the truck need to be taken with both doors closed. You don't want the new seat to interfere with the truck's armrests. It is a good idea to allow an inch of clearance on both sides of the new seat.

The older the truck, the more likely you will have leg room limitations. These trucks relied on very upright driving positions to provide what leg room was available. The stock backrests tended to be thin in a compromise of leg room over comfort. The luxury seats are inevitably much thicker. This thickness provides comfort, but it also pushes the driver closer to the dashboard. If you are tall, you will want to find a luxury seat with the thinnest possible backrest.

Power Seats

Power seats are a nice option to consider when swapping. The power bases usually come with the luxury seats discussed for trucks, but you can also adapt these mechanisms to other seats. When you purchase power seat bases from a wrecking yard, be sure to get all the related parts. Get the wiring harness, relay box, cables, and mounting hardware. Don't forget the all important switch panel from the side of the seat. You will need to mount this switch panel to your seat or the seat base. When you hook up the power seat motor to a power source, it is a good idea to include a 20-amp inline fuse. ■

Dyes & Color Changes 13

A common problem when buying a used car or truck is not finding your first choice of exterior and interior colors. Changing the exterior can be a time-consuming and expensive proposition. Changing the color of the interior is easier, but doing it correctly requires some work and attention to cleanliness. Too many people have an incorrect notion that changing interior colors is just a matter of getting a couple aerosol cans of vinyl dye and spraying the seats and door panels. As many people have learned the hard way, a quickie dye job won't last much longer on your interior than it will on your hair.

There are places where dyeing upholstery material is your only affordable solution. Many reproduction interior parts are only available in a single color, usually black. This is great if you have a black interior, but if your car is another color, you will need to dye the new parts to match the original color scheme. The problem of non-matching colors is also something you can encounter with used parts. It is difficult enough to

When you're trying to build a car on a budget, sometimes what kind of deals you find can determine the direction of the interior. The door panels and rear quarter panels were shot on this '57 Chevy, but a very decent set of mildly modified black panels were found at a swap meet.

Dyes & Color Changes

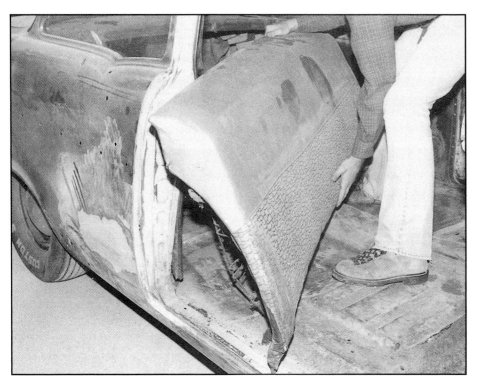

The silver colored seats in the Chevy were pretty sound and definitely usable. It was decided that changing their color would be the most economical way to improve the interior. Seats should be out of the vehicle for a color change.

find a nice used dashpad or set of door panels in a wrecking yard (or at a swap meet), but when you throw in a particular color requirement—good luck. With used interior parts, odd-colored parts are often considerably less expensive than popular colors. The choice comes down to living with your correct color, but damaged original parts, or dyeing the better condition replacement part.

If you are having trouble finding the right color replacement parts, consider dyeing as a temporary solution. You can always upgrade your interior when you find a perfect set of new old-stock parts or when the aftermarket industry decides to reproduce the parts in question. If the dyed parts are still in good condition, you can probably sell them to someone trying to build their car on a budget.

Dyeing can help solve small problems, but changing the color of an entire interior can be a big undertaking. The general rule is that the larger the area to be dyed, the bigger the potential problems. This is not to say that you can't do this job at home; it is just that satisfactory, long lasting results are hard to obtain.

Which Parts Are Best?

There are areas that respond better to dyeing than others. A key factor is how much wear the area will receive. Items like seats aren't the best choice for color changes, while parts like kick panels or rear side panels are pretty safe. As I mentioned before, sometimes you don't have any choice other than dyeing. If you have a choice, include the wear factor along with items such as cost and part availability.

Interior parts such as seats involve a fair amount of movement on the portions of the vinyl that make contact with passengers. Since many dye jobs really only cover the surface of the material, movement can lead to cracking and chipping. More static interior parts such as door panels have very little chance of cracking, although these parts are susceptible to scratches.

Dye Quality

The quality of the dye is an important factor in how well the dye penetrates or sticks to the vinyl. People have been known to simply paint door panels with common aerosol paint, but this type of paint only lies on the surface of the vinyl. This technique is strictly a down and dirty, "sell a junker in a hurry" trick.

Carpet Dyes—Some vinyl dye products are also marketed as carpet dyes. Carpets can be dyed, but finding replacement carpet sets is much easier than other interior parts, so many people go with new carpet. A problem with dyeing carpets is achieving full penetration. This is more difficult with the deeper styles of carpeting. It helps to apply a couple light coats of carpet dye. Use a stiff bristle brush (a common household scrub brush will work) to help work the dye into the carpet. The brush is also used to keep the carpet fibers separated while drying. You don't want to apply too much dye and end up with clumps of dyed carpeting. The dyes can make the carpeting stiff, so use the scrub brush to loosen the fibers after the dye has thoroughly dried.

COLOR MATCHING

Black interiors are pretty simple to match, but different sections of an interior can fade and age at different

Dyes & Color Changes

rates, leaving you with pieces that don't match. Even basic colors like white can fade unevenly, depending on where it is located in the car and whether it is exposed directly to the sun.

Touch Up

If you want to touch up a few parts of your interior, you need to decide what shade will be the key color. If you are dyeing different colored components to match your existing parts, the new dye will need to match properly. Automotive paint supply stores can assist you in obtaining the best possible color matches. Their reference books should have the codes for most vehicles. If they don't have the codes or if you want to match colors with your existing (but slightly faded) parts, the paint store can custom match the dye to your needs.

A good way to match the original interior color is to bring a section of the interior to the paint store. Try to pick a component that is in the best possible condition. Then show the paint store the back side of the piece. An area like the back side of a door panel where the material is wrapped around the panel board should be unfaded. On a part like a door panel, match the dye to the bottom of the panel. Another possible place for unfaded vinyl is the hidden part of an armrest.

Some paint stores like to have both the factory color code and a good sample section of the vinyl. This way they can start with the factory code and custom match the dye to your actual sample.

An important aspect of color matching is to buy ample dye at the time you have it mixed. Dye batches are seldom exactly the same and

There are several places to get upholstery dyes. Local auto part stores usually carry a limited selection, but reproduction interior companies will have the correct colors for the models of cars they represent. Automotive paint supply stores also carry vinyl dyes, but these dyes must be sprayed on with a spray gun.

custom matching a previously custom-matched color can be difficult. To avoid having two different shades of the same color, be sure to order more than enough dye. It is far better to have some leftover dye than to run short. Having extra dye can come in handy if you need to do some touchup work at a later date.

SURFACE PREPARATION

Cleanliness can't be emphasized enough when it comes to obtaining a good, long-lasting dye job. Dyes won't adhere well to poorly prepped material. Automotive interiors absorb a great deal of dirt, grease, and assorted chemicals. The porous nature of upholstery means that these contaminants can be very well imbedded. Removing them requires thorough and repeated cleaning. Even if you are the original owner of the vehicle and have taken excellent care of it, there can still be a lot of dirt present in the upholstery. Used parts from a wrecking yard are often grime central, so you almost can't get these parts too clean.

It is best to use all the products from the same company to ensure chemical compatibility. Since many interior panels have some type of texture or grain, you will need to scrub extra hard to remove dirt from these surfaces.

Dyes & Color Changes

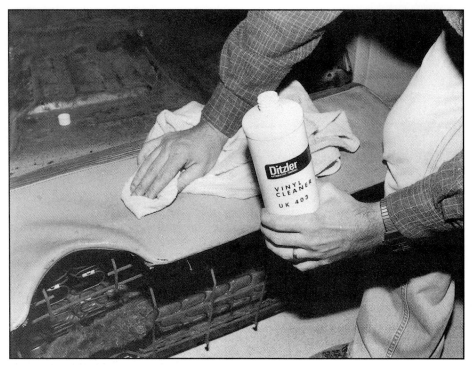

The key to a good color change is surface preparation. The seats or door panels must be thoroughly cleaned. A lot of grease and silicone can get on upholstery and they make it hard for paint to adhere properly. Use a cleaner that is compatible with the dye. Follow the manufacturer's directions to the letter.

When you lift up dirt from textured areas, you also need to remove the dirt. Follow the scrub brush with a damp cloth to wipe up the cleaner solution and the dirt. Turn the cloth frequently so that you blot more dirt, rather than put the dirt back into the panel.

Corners, seams, and crevices can be particularly tough to clean. Take the effort to clean these areas. Otherwise, you can end up with poor dye adhesion. Old toothbrushes work well for cleaning hard-to-reach areas. These cleaning steps should be repeated two or three times. It is best to clean small areas instead of large ones. This way you are more likely to do a thorough job. Don't forget to clean all the edges (and the part that rolls over to the back side) of items like door panels. You want the dye to adhere all around the edges.

Vinyl Protectants

If you have used vinyl protectants in the past, all traces must be removed. If you have just purchased the car, chances are these protectants were used at one time or another. Most of these protectants contain silicone, which is death to dye and paint jobs. The silicone can cause the vinyl dye to "fisheye" or leave little areas where the dye won't properly adhere.

Some product systems suggest rinsing the upholstery with water. However you do it, you must remove all traces of the cleaner after the dirt is gone. Be careful not to leave lint on the surfaces, either. A lint-free rag is best for the final wipe down. The upholstery should be thoroughly dried after cleaning. They can either be dried with compressed air, or just let the parts set until dry. Compressed air is good for removing moisture from crevices, such as near piping trim.

Even new reproduction parts that are going to be dyed should be thoroughly cleaned. These parts can be coated with release agents such as silicone, which must be removed.

Once you have the parts cleaned, try to minimize your hand contact. You can have a lot of oils and other contaminants on your hands. Wearing disposable latex gloves is a good way to keep the upholstery extra clean.

Most dye systems use a vinyl conditioner after the cleaning process. Follow the directions for the application of the conditioner. The conditioner helps the vinyl accept the dye. Some companies offer different cleaners and conditioners for different materials. Be sure to get the right products for the job. Consult your local paint supply store for their recommendations.

COLOR APPLICATION

Upholstery dyes come in aerosol cans and bulk. Professionals rarely use the spray cans. You can get custom mixed and exact match colors in the bulk dyes. Bulk dyes are best for covering large areas. Spray can vinyl dyes are best for small jobs or touching up your existing upholstery.

A touch-up or detail spray gun is most often used for applying vinyl dyes. These guns are easy to handle if you are applying any color inside the vehicle.

Flex Agent

Some color change systems use a flex agent in the dye. Flex agents are mostly used on seats and dashpads. The flex agent makes the paint more pliable and less apt to crack or chip.

The dye should be applied in several light coats, rather than one

Dyes & Color Changes

This seat wasn't really sprayed with the can of vinyl color next to the spray gun. A touch-up gun (shown) is ideal for applying the color coats. It will hold enough paint to cover a seat without refilling and it is light enough and small enough to maneuver easily.

heavy one. You don't want the paint to run. Don't try to cover the underlying color with a single application of the new color. Many experts suggest applying the coats in a perpendicular manner. That is, apply the first coat horizontally and follow it with a vertical coat.

In general, the touch-up gun should be held about a foot away from the surface. The air pressure should be 35-45 psi. Keep the spray gun moving to avoid runs. Allow about ten minutes between coats for flash time. It is best to spray the parts at room temperature. If the temperature is too cool, you will have an increased risk of runs.

Besides runs, heavy applications can cover up any texture in items like kick panels. Watch the condition of the texture as you apply each coat of color. You want just enough color to cover the underlying one, but not so much dye as to alter the texture. If you are unsure about coverage, use a strong light to examine the freshly painted surfaces.

High humidity or too high temperatures can lead to blushing or a hazy finish. If this happens, try applying a light coat or holding the spray gun a little farther away from the surface. When the spray gun is held back, the paint will be drier when it hits the surface. You want it dry enough to combat the blushing, but not so dry that the finish is rough or grainy. Light coats are the key to avoiding blushing.

The final coat of color should be allowed ample drying time before the parts are handled. If the dye isn't dry, your fingerprints will show. Some dyes are dry in about an hour, but it is best to wait a couple hours. If you can wait overnight before handling the parts, that is even better.

If you are using an aerosol can, let it sit in some warm water before using it. This will help the atomization of the paint. Shake the can very well before applying the vinyl dye. Be careful with spray cans when you get near the bottom of the can. It is better to use a fresh can than to risk getting splatter from the bottom of the can. If any paint builds up on the nozzle, clean it off for the best possible spray pattern. ■

Minor Repairs & Detailing 14

The focus of this book is restoring interiors and installing new parts, but we will also cover some minor repairs and detailing. Sometimes a simple repair will give new life to an interior component. Other repairs may not be as good as a new part, but they can get you through until you have the time and/or money to replace the worn parts.

Detailing is an automotive term that means going beyond just a simple cleaning. In other words, when you detail a vehicle, you pay careful attention to all of the little details, from minor rips and tears, to decals, polishing, deep cleaning, etc.

There are complete books on the subject of detailing so this chapter will just give a general overview. If you have installed fresh upholstery and carpeting, it is much easier to keep it clean than it is to try rescuing a dirty interior. A dirty interior won't last anywhere near as long as a clean interior.

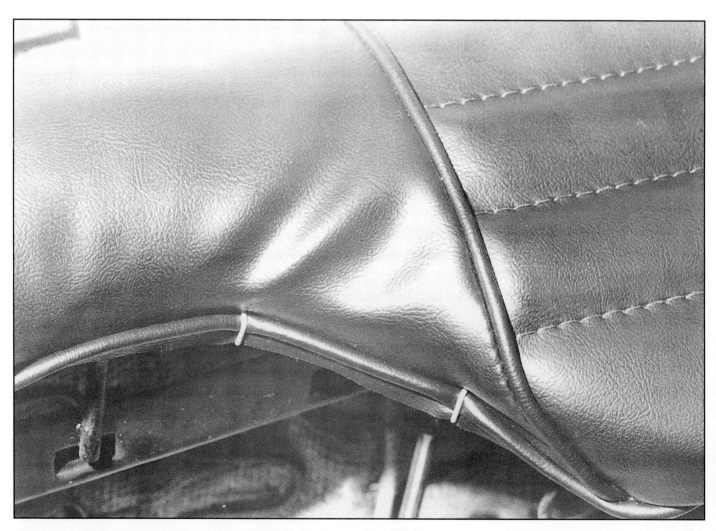

The more you learn about automotive upholstery, the more you will notice the differences between a great job and one that is merely passable. The fit of this seat cover where it curves over the driveshaft tunnel was noticeably wrinkled.

MINOR REPAIRS & DETAILING

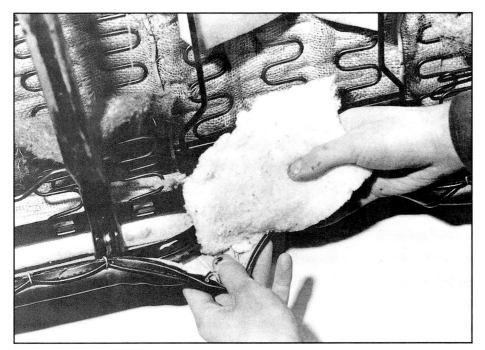

Part of the reason for the wrinkles was a lack of padding in the area. Padding can get compressed over time, but reproduction seat covers are made to the same dimensions as if the underlying parts were in perfect shape. The hog rings were removed in this area so some new cotton batting could be inserted.

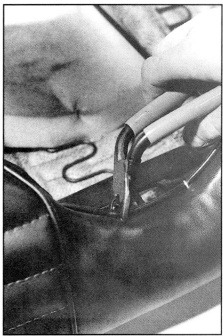

The cotton batting was smoothed out to fill the voids in the seat cover. Then the cover was tightly stretched over the seat frame and secured with new hog rings.

MINOR REPAIRS

Complicated repairs need to be handled by professionals, but you can handle the minor problems. You can also do some of the preliminary work on major repairs and save money. For example, if a seat cover has a ripped seam, you can save money by removing the seat cover and taking it to the upholstery shop for the actual sewing. That way you will only pay for a small amount of the shop's time. The techniques for removing and reinstalling the seat cover are detailed in Chapter 8.

Check with the upholstery shop before you take your seat apart. There may be a reason why the shop would want to see the whole seat (besides running up your bill). If the problem is more than a couple inch tear, or if a large piece of upholstery is missing, the shop may need to work with the whole seat. Some shops aren't cooperative about only doing partial jobs, but by making a few calls, you should be able to find a reasonable shop.

You might also want to visit an upholstery shop before you begin repairs to get an estimate and their professional opinion. Some rips may be tough to repair without the repair being very obvious. You may also find yourself in a situation where the repair is almost as costly as installing new seat covers. If the estimate is free, avail yourself of their expert opinions.

In the case of a ripped seat cover, the professional shop may also point out that the foam is badly worn underneath. If you simply sew up the rip, you could still have a less-than-perfect seat because of the deteriorated foam base. Building up the foam base is covered in the seat upholstery chapter.

The professional shop may also point out that certain kinds of upholstery damage are nearly impossible to repair. For example, some seat covers look like they have sewn-in pleats but they are actually

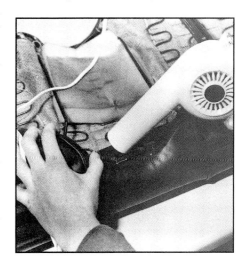

To make the area even smoother, a hair dryer was used to carefully warm the vinyl. The warm vinyl was smoothed with thumb pressure. The heat helps the vinyl relax, but too much heat can do more damage than good.

Minor Repairs & Detailing

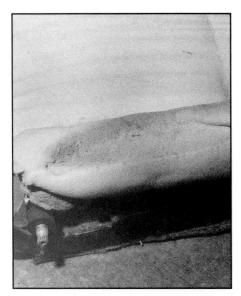

The outside edge of seat cushions (especially the lower section of the driver's seat) gets a lot of use from sitting and sliding in and out of the vehicle. If the foam is too compressed (or coming apart) the new seat cover won't fit properly. Also, an old cover that is otherwise in good shape will sag and not look right with poor underlying foam.

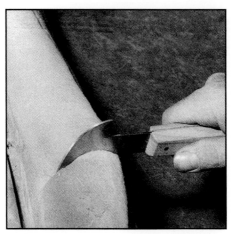

The damaged section of foam needs to be cut out. Cut out all the of the worn out foam until you hit good foam. A trimmer's knife will cut the foam. Go slowly so you don't remove too much foam.

pressed into the material. There isn't an actual seam to sew. You can "mend" the rip, but it will look more like an old pair of jeans than a new seat cover.

Rips can be in areas of high use and stress. Sometimes the material is actually worn out as well as ripped. In these cases the shop will probably recommend a new seat cover.

The main thing to remember about repairing upholstery is the substantial difference between repair and restore. The main focus of this book is restoring your upholstery to like-new condition. Upholstery repairs are more of a temporary measure until you have the time and/or money to do the job right. Repairs can come in handy when you want to sell a vehicle with a value that doesn't warrant a full-blown restoration. A seat cover that has had some seams repaired might not be as nice as a brand-new cover, but it won't call the negative attention that frayed seams will. People don't expect ripped seats when they buy a used car. They will deduct (or pass on the car altogether) money for ripped seats, but they might not pay you more for brand-new seat covers. In cases like this, repairs can be an excellent investment.

Keeping with the theme that repairs seldom are as good as restoration work, don't expect perfection from repairs. Your goal should be to make the repair as unobtrusive as possible. You don't want the repair to attract more attention than the damage.

SEATS

As mentioned previously, you first need to determine if the seat damage is something that is feasible to repair. The nature and location of the damage will often determine the ease of repair. Ripped real seams (as opposed to formed or pressed vinyl) are pretty easy to sew. If you have cloth upholstery that is so worn that the weaving is separating, that can be difficult to repair. You may be able to replace a whole panel, but the replacement panel may look too nice compared to the rest of the seat if it is quite worn.

In most cases, I suggest removing the seat cover and having a professional shop sew the damaged seams. It is possible to repair a ripped seam while the seat cover is still on the seat frame. There are curved needles designed for this kind of repair. The biggest problem with sewing while the seat cover is in place is holding the material on either side of the rip tightly together while the sewing is performed.

The sooner you can catch a rip, the easier it will be to repair. If there are loose threads, knot them so that they don't come further undone. Once a rip starts, it is much easier to enlarge it. The principle is the same as a rip in your jeans.

Foam Patches

If you remove the seat cover for sewing, check the condition of the underlying foam. Areas that have rips often have poor foam as well. If the damage isn't too bad you can glue in a patch piece. The biggest problem with patching foam is getting it nice and smooth. You don't want a big lump where the foam patch was installed.

You can make the foam patch relatively smooth by trimming the foam to the same height as the surrounding foam. Professional trimmers use special foam knives to shape and shave the foam. You can carefully trim the foam with a single edged razor blade or even an old (not the Sunday dinner set) serrated steak knife.

Another trick for making foam repairs smooth is to cover the area with a piece of lightweight cotton or

MINOR REPAIRS & DETAILING

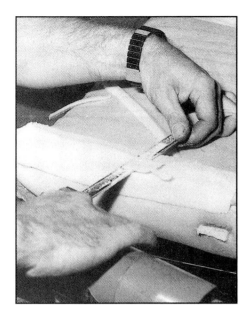

There are electric foam shaping tools that professionals use, but an old electric carving knife will do a similar job. You can also cut out the bad foam with a serrated steak knife. When shaping foam, remove a little at a time.

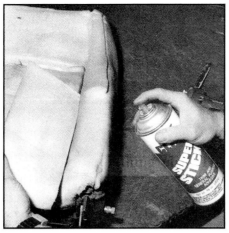

Rather than try to cut out a perfect patch section of foam, you can glue a larger piece to the cushion and then trim it to size. Trim adhesive should be applied to the patch and the original seat cushion.

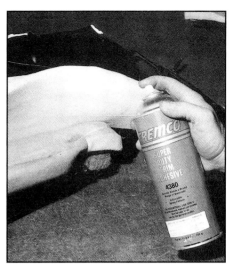

Seats with compressed side bolsters (as opposed to sections of deteriorated foam) can be easily built up with pieces of half-inch foam. The new foam should be wrapped all the way around the bolster. It should run the full length of the bolster for a smooth, uniform look. Secure the foam with trim adhesive. After the glue has set, any excess foam can be trimmed.

muslin material. By stretching the material tightly, the foam repair will be flat with the rest of the seat. If the actual seat cover is made out of sturdy vinyl you won't have as much trouble with foam repairs as if the cover is velour or some other fabric.

Vinyl Repair Kits

While most repairs to vinyl seats should be done with a patch panel, there are vinyl repair kits for fixing simple rips. These kits can be useful for short term repairs for rips that aren't along seams. The basic idea behind these kits is gluing the two sections of vinyl back together. Some of these vinyl repair kits use heat (like a modified wood engraving tool) to bond the damaged areas, while others just rely on the strength of the glue.

Procedure—A few hints when using a vinyl repair kit are keeping the vinyl edges tight, getting a perfect match up of the torn edges, and using the smallest amount of glue that will accomplish the bond. Duct tape or good quality masking tape can be used to keep the two sections of vinyl together until the glue sets. When using the tape, leave ample room between the strips for the glue. The better the edges are aligned, the smoother the repair will be. You don't want a build up of excess glue, so apply the glue sparingly. You will need to remove the tape and glue those areas of the rip after the first repairs have cured. These vinyl repair kits also can be used on door panels and dashpads.

Holes

Burn holes like those caused by cigarettes usually involve removing the seat cover and patching the area. The size of the patch depends on the design of the seat cover. You will usually need to go to the nearest seam. I have seen otherwise perfect velour seats ruined by a cigarette burn hole in the center of the cover. The velour isn't supposed to burn, but the hot ashes leave kind of a melted hole.

There are inexpensive velour repair kits for automotive upholstery available. These kits come with velour fibers that are inserted in the burn holes. The kits come with basic color selections and hopefully one of them will be close enough for your seat. Since these velour repair kits are so reasonably priced, you could try one first. Then, if you're not satisfied with the repair, you can take the cover off and have it fixed at an upholstery shop.

Fading

It is not uncommon for seat upholstery to fade with time. Depending on where the car was parked and under what conditions, you could have a situation where parts of the upholstery were faded while other parts were not. The first thing you can do with fading is thoroughly clean the material. If you have an easy to match upholstery color, you could try some vinyl or fabric dye.

125

Minor Repairs & Detailing

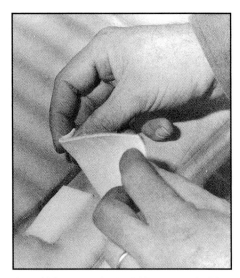

A small foam patch should be covered with a slightly larger section of cloth. You don't need to cover the full length of the bolster unless it is excessively compressed. The cloth ensures the outline of the repair won't show through the seat cover.

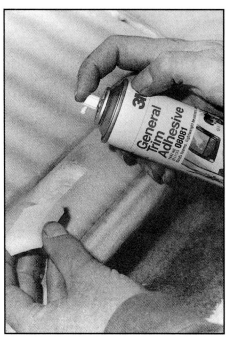

Sometimes the seat cushion will have a small puncture or divot. The area can be cut out and fitted with a block of new foam. Trim adhesive is used to bond the two pieces of foam.

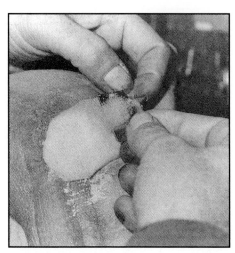

Small foam repairs can be trimmed with a single edge razor blade. Use a fresh blade so that it is sharp. Be careful not to cut yourself.

Information on dyeing is in Chapter 13.

CARPETS

Carpets can have repair problems similar to those encountered with seats, except that rips are less likely. Replacement carpet sets are so affordable for most vehicles that I suggest replacement over repairs in all but the most minor situations. Since replacement carpet sets are reasonably priced and easy to install, you don't have much to lose if you do try to make some carpet repairs.

If by some means your carpet does get ripped (e.g., the cargo area of a van or station wagon where some sharp object snagged the carpet as it was loaded or unloaded), you can remove the carpet and tape the backside of the tear. You could also join the two sections with the same type of glue used for vinyl repairs.

Cigarette burns are a common carpet problem. Unless you have some expensive German woven carpet (the kind often used in Mercedes, BMWs, and Porsches) you can replace the damaged area with a plug of good carpet. The process is sort of like hair plugs for your carpet.

Carpet Plugs

To make carpet plugs, look under the seat or under the sill plates (if you remove any carpet from under the sill plates, be careful not to remove too much so that there is still enough carpet to tuck under the sill plate) for some carpet that is in good condition. Under the seat is usually the best source. Before you remove any carpet, slide the seat all the way forward and all the way back to be sure you aren't removing an area that shows.

You can use an X-Acto knife or a single edged razor blade to cut the carpet. When you cut out the burned or permanently stained spot, try not to cut any more than absolutely necessary. This plug repair works best on relatively small holes. Fill the hole by gluing the plug of good carpet in the hole. After the glue has dried, you can use your hand or a carpet brush to gently work the area so that the plug blends with the nap of the carpet.

DASHBOARDS AND SIDE PANELS

Heat and ultraviolet rays are a dashboard's worst nightmares. As silly as some of those windshield screens look, they do help preserve dashboards as well as reduce overall interior temperatures. Cracked dashpads are the usual result of heat and sun damage. If the top of the dash looks like a spider web, or has crevices instead of cracks, it is beyond repair. The best solution is to replace the dashpad, which is covered in Chapter 7. However, minor cracks and tears can be repaired with a dashpad repair kit, commonly found at auto parts stores and available from the Eastwood Company.

MINOR REPAIRS & DETAILING

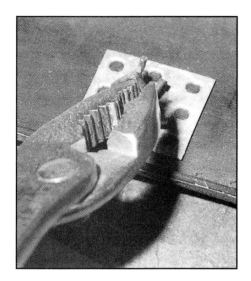

Door panels that have been repeatedly installed and removed can deform the retainer clips. If the clip is collapsed, the panel won't fit tight. New clips are the best solution, but you can save most clips with a little tweaking with a pair of pliers.

There are a variety of vinyl and dashboard repair kits available. Some are little more than specialized glue, while other kits include filler, glue, a heat tool, grain duplicators, and various colored dyes. None of these kits are particularly expensive. The type of repair kit that you use depends on whether you have a simple crack or a divot that needs to be filled.

Repairing Cracks

In cases where the underlying foam has deteriorated or a chunk of it has fallen out, some type of filler needs to be used to bring the area up to the level of the rest of the dashpad. You can use the filler that comes with the kit. Bigger holes can be filled with expandable spray foam (the type found at home improvement centers for weather sealing houses). The foam hardens into a firm urethane that can be trimmed or sanded to match the dashpad's surface.

After the crack or hole had been built up to the proper height, the topcoat material is applied. Sometimes a graining material will have to be used to match the grain of the dashpad. The repair needs to be painted. You can use the paint in the kit (the better kits have color charts and directions for matching specific colors) or repaint the entire dashpad. Since cracked dashpads are often faded, it isn't a bad idea to repaint the whole unit.

Plastic Parts

Another dashboard problem is broken or cracked plastic components. Items like gauge clusters, switch mounts, and vent grilles can get cracked or otherwise damaged. If you can't find a good replacement item, you can repair most problems with glue. Epoxy or special plastic glues should be tested on some old unwanted part first. Some glues can "melt" plastic.

The best way to repair these problems is from the back side. If the part has a plastic chrome finish you will probably need to remove the chrome where the glue goes. Before you damage a rare dashboard component consider practicing on an old plastic dash component. You should be able to get such a broken part at your local wrecking yard for next to nothing.

Touch Up—You will probably need to repaint or at least touch-up the paint on the front side of the plastic dash part. If your part has plastic chrome plating there are companies that can rechrome the parts. You might be able to find such a company in a big city, but they may or may not be interested in doing small, odd jobs. Your best bet is to check the "Restoration Services" section in *Hemmings Motor News* for companies that specialize in plastic

Windows that are difficult to roll up and down are often the result of window regulators that have never been serviced. When the door panels are off, it is pretty easy to access the regulators. About four fasteners hold the regulator to the door skin. When they are removed, the upper rollers need to be disengaged from the track at the bottom of the glass. The regulators go in and out of a door access panel.

chrome plating for cars.

Often a faded or scratched plastic dash component can be successfully touched up with simple model paint and a small brush. The key is to use a very small brush (like a #001 brush) and a minimal amount of paint. Another trick is to apply the paint with a toothpick. The toothpick technique is especially useful when touching up the raised letters on dash panels. Hobby shops also have paint pens which are great for touch-up work. While items like gauge clusters are apart, clean the plastic lenses and apply plastic polish. This will greatly improve the looks of a faded gauge cluster.

Side Panels

The techniques used to repair rips and cracks in seats and dashpads can be used for most side panel repairs. If the side panel has a carpeted lower section, the carpet repair techniques

127

Minor Repairs & Detailing

All the areas that roll or slide in the window regulator should be cleaned and coated with white grease. If the nylon rollers are deteriorated or if they have flat spots, replacement rollers are available.

Seat adjuster tracks will benefit from a good cleaning, followed by an application of white grease. When working with the seat tracks, mark the proper location of the springs.

can be used, although it could be easier to replace the whole section of carpet.

Damaged or worn armrests are a common problem. Replacement pads are the easy way out, but you can also recover the pads. Recovering armrests is explained in Chapter 6.

Badly scuffed side panels are common. A thorough cleaning followed by touch-up vinyl spray should greatly improve their looks. Remember that the better the cleaning job, the better the paint will adhere. The vinyl color spray should be applied in a couple light coats, rather than one heavy coat.

The bottom edges and the lower sections in general can get worn or damaged on cars that have panel board for backing (as opposed to the more modern, molded plastic side panels). Again, replacing the whole side panel is the optimum solution, but repairs can be made. If the deterioration is caused by water, you should try to locate the cause and fix it.

You can get new panel boards for many popular cars. If you can carefully remove the outer covering (not always feasible), you can reinstall the covering on the new panel board. Most likely, only the bottom few inches will be damaged. You can peel back the upholstery and cut off the damaged section. Use a straight edge for an even cut. Make a new replacement section out of waterproof panel board that is the same thickness as the original board.

Butt the new section up to the old panel board. Use good quality duct tape on both sides of the seam. This won't be as strong as a single piece of panel board, but it will work to hold the shape of the outer upholstery. Remember to make the properly aligned holes for the mounting clips. Re-glue the upholstery to the patched panel board section and re-attach the entire side panel to the door.

If some of the retaining clips are loose or missing from a door side panel, the panel can be better secured with strips of Velcro. Get the Velcro with self-adhesive backing. This technique also works well to keep carpet in place.

STEERING WHEELS

The most common problem with steering wheels is cracks in the plastic rim. These restoration techniques are covered in Chapter 11.

Aftermarket steering wheels sometimes work their way loose from the steering hub. If you encounter any wobble or looseness, check the retaining nut.

If your steering wheel has a padded center section, it can fade with time. The vinyl can be resprayed using the same techniques as for seats, side panels, and dashboards.

Horns

Horns can be faulty. The problems are usually related to poor or dirty electrical contacts, but the first thing you should check is the fuse. Examine the electrical contacts, both in the horn button and at the horns

MINOR REPAIRS & DETAILING

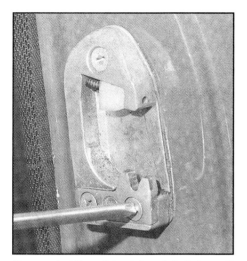

Door post striker plates take a lot of abuse. If they aren't adjusted properly (and the door hinges also need to be adjusted), closing the door can be difficult. You can get replacement striker plates for many vehicles.

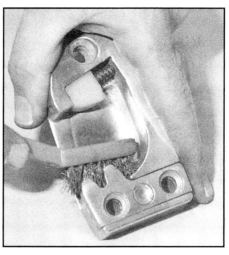

The striker plates tend to accumulate a lot of grease and grime in the corners. A small brass bristle detailing brush works well to clean the striker plate. A minimal amount of polishing with Wenol all-purpose metal polish will have it shining like new. Wenol works very well on all kinds of interior trim items.

themselves. You can run a jumper wire from the positive terminal of the battery to check the horn operation without going through the horn button. Remove the horn's electrical lead wire and place the jumper wire here. If the horn doesn't blow, the problem is either the ground connection or the horn. If the horn is OK, the problem is either in the horn button contacts, the wires leading from the steering wheel to the horn, or the horn relay. If you think you have a problem in the wiring system, consult a factory shop manual for specific details relating to your vehicle.

DOORS

Window crank mechanisms, door latches, and power interior accessories can all wear with time. The best time to service any door related components is while the door panels are off for replacement or restoration. Often, simple cleaning and lubricating will improve the condition of window regulators and lock mechanisms. There can be nylon rollers to ease the operation of windows. Check to see that any rollers or bushings aren't worn out or flat-spotted. Replacement parts are usually very inexpensive.

Check Alignment

It is also possible that the various inner door parts are out of alignment. There are usually adjustment features. Check a factory body manual for specific details on your car or truck. Door latches can be out of alignment due to sagging door hinges. The hinges can be replaced or rebuilt. Often the hinge bushings wear out. Door hinges could have been improperly installed after a damaged door was replaced.

Check Drain Holes

While you have the door apart for maintenance or repairs, be sure to check the drainage holes. Plugged drain holes lead to rust problems. If your windows seem to be loose or rattle too much, the window whiskers are probably worn out. Replacement whiskers will greatly improve the seal.

Power Accessories

Problems with items like power windows, power seats, and remote mirrors are most often electrical ones. A factory shop manual and an understanding of troubleshooting techniques with an ohm meter will go a long way toward solving these problems. Sometimes the problems can be as simple as a blown fuse. It is also a good idea to check the mechanical aspects of the option. Junk under the seat could hamper the operation of power seats. Power window and seat tracks can benefit from periodic lubrication. Some newer vehicles have plastic or nylon teeth on the power window tracks. These tracks can break or a couple teeth can be missing. When working on power windows, you will often need to remove the glass from the regulator channel. Be careful not to

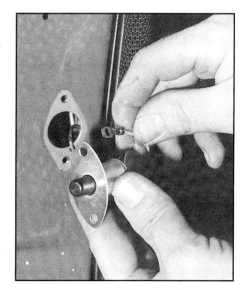

When you restore a vehicle's interior, you are likely to find lots of annoying little problems such as defective door jamb light switches. New switches are inexpensive and easy to install.

129

Minor Repairs & Detailing

White lithium grease is available in aerosol cans. With the long applicator nozzle, you can lubricate many moving parts (like this inner door handle pivot) without removing them from the car.

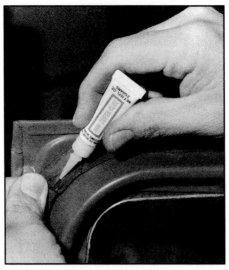

Sometimes the weatherstripping around doors can develop a few minor tears that aren't bad enough to warrant replacing the whole piece. These tears can be repaired with an application of Super Glue.

If you need to install new weatherstripping or if a section of the old weatherstripping comes loose, a small block of wood works well to push the mounting pins into the door. The wood disperses the pressure, unlike a screwdriver which can puncture the rubber.

drop and break the glass. Most electrical connectors are designed to only fit one way. Don't force the connectors if they don't go on easily. When working on power seats be careful not to pinch your fingers in the power mechanisms.

Repair Weatherstrip

Door related wind problems are often related to weatherstrip problems. Badly worn weatherstripping should be replaced. If you have a few minor cracks or tears in the weatherstripping they can be repaired with Super Glue. Apply a couple drops of glue to both sides of the tear and hold the weatherstripping in place until the glues sets up. If you are installing new weatherstripping a tip is wipe the weatherstripping with solvent (such as wax and grease remover) before putting on any adhesive. The mold release that the factory uses on the weatherstripping can reduce the effectiveness of the adhesive.

DETAILING TIPS

By the time you have reached this part of the book, hopefully your interior is in like-new condition. Maintaining a restored interior is pretty easy as long as you clean it frequently. It is much easier to do a weekly or bi-weekly quick cleanup than to wait months and have to do a major cleaning.

Vacuum Frequently

The single most important thing you can do to keep your interior clean and as new-looking as possible is to vacuum frequently—at least once a week. Use a plastic crevice attachment in a back and forth motion to get up as much dirt as possible. Stubborn pine needles and burrs can be dislodged with a plastic-tipped hair brush or a scrub brush.

The seats and the rest of the interior can be vacuumed with a combination of the crevice tool and a small brush attachment. Press down into the corners and crevices of the seats as you vacuum them.

There are special detailing brushes for getting dust out of tight areas, but you can also use an old toothbrush. For larger areas you can use an inexpensive paint brush. Cut the bristles to about half their normal length. This will make the bristles stiffer. Many detailers also use cut down paint brushes for applying shampoo and working it into the seams and crevices of the seats and door panels.

Carpets

After the vacuuming is done, it is time to clean the carpets. There are a variety of automotive and household cleaning products such as Resolve® that will do the job. Some people rent household carpet steam cleaners (the kind that they rent at grocery stores), but they seem like a lot of equipment for a relatively small amount of carpet. You could be a hero and shampoo the household carpets at the same time you do the car.

Minor Repairs & Detailing

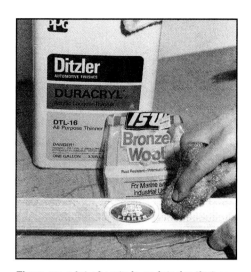

There are a lot of parts in an interior that can be rejuvenated with a good cleaning. You can save money now and upgrade the parts later if you want. Door sill plates take a lot of abuse, but cleaning them with lacquer thinner and lightly polishing them with bronze wool will greatly improve their appearance.

Use a scrub brush to work the shampoo or cleaner into all areas of the carpet. Change the scrubbing pattern to assure a thorough cleaning. If the carpet is exceptionally dirty, you may have to apply the cleaner more than once. Rinse the carpet thoroughly after cleaning. If you can dry the carpet outside, do so. If your shop vacuum is a wet/dry model you can remove most of the excess moisture with the vacuum.

It is important to thoroughly dry the carpet after cleaning. If you don't, the carpet will develop mold and mildew, and the odor will permeate throughout the car forever. Wet carpet can also lead to floorpan rust problems.

Mold—If your car has a wet, moldy smell, try to find the source of the problem rather than just covering it up. The biggest causes are water leaks. A leaking heater core is a common source of unwanted interior moisture. After you shampoo the carpets you can help eliminate the moldy smell by placing a couple of small containers of baking soda under the seat. Plastic margarine tubs work well for holding the baking soda. Replace the baking soda with fresh baking soda every few weeks until the odor is gone.

There are also "dry" carpet cleaning products. These products are usually sprayed on the carpet, rubbed into the fibers, and allowed to dry. The dry residue is then removed with a regular vacuum cleaner.

Seats

How you clean the seats depends on the type of material used to cover them. Vinyl seats are the easiest to clean. Leather is reasonably easy, but you need to use products designed especially for it, such as Lexol®. Cloth seat cleaning is similar to carpet cleaning, but you must be careful not to soak the padding underneath.

A variety of cleaners will work on seats, but it is a good idea to first test the product on an inconspicuous part of the seat. You want a product that is strong enough to clean, but not so harsh that it harms the material. Some products can be fine on the actual vinyl, but destructive to the thread used on the seams.

Vinyl—Vinyl upholstery is pretty durable so you don't need to be as particular about the type of cleaner you use on vinyl. It is always a good idea to use a product specifically designed for vinyl, but if you don't have any, a general cleaner or even a mild household cleaner will work. If the vinyl upholstery is more dusty than dirty, you can simply wipe it down with a damp cloth. Really dirty vinyl can benefit from light scrubbing with a soft scrub brush. You want a brush that will loosen the dirt without abrading the vinyl.

You don't want the soap to dry on the vinyl so rinse as needed to avoid dried soap residue. A damp sponge is a good way to rinse away the cleaner without using too much water. If you want to keep the seats as dry as possible while you are washing and rinsing them, keep a wet/dry shop vacuum running in one hand while you use the sponge with the other hand. A crevice tool works well for this purpose.

If you are cleaning the seats while they are still in the vehicle, consider putting some old towels around the base of the seat. The towels will help keep the dirty residue and rinse water away from the carpets. This is a reason why it is so much easier to do a thorough cleaning with the seats removed from the car.

Cloth—Cleaning cloth seats takes more care than vinyl. You need to be careful not to soak the fabric. There are special cleaners for cloth seats. Many of these cleaners generate more foam than moisture. Some cleaners are the "dry" type similar to carpet cleaners. A wet/dry shop vacuum is a good idea to keep moisture saturation

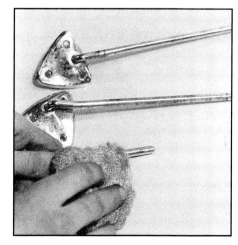

Minor oxidation can make chrome plated or stainless steel trim pieces look shabby. Items like these sun visor arms respond well to a light rubbing with bronze wool. An application of metal polish completes the job.

Minor Repairs & Detailing

If you have a rear view mirror that won't shine up with chrome polish, you can get some more use out of it by painting the back of the mirror black. The surface needs to be cleaned and then scuffed with a Scotchbrite pad. A little surface "tooth" helps the paint adhere. Prime the mirror before applying the black paint. Be sure to mask all areas that shouldn't be painted.

to a minimum when cleaning cloth upholstery. If you are unsure about how to clean cloth upholstery, consider having a professional detail shop do the job. It is better to pay a few dollars for a competent cleaning than ruin the seats yourself.

Leather—Not too many cars have real leather upholstery, but those that do require special cleaning products. These leather care products include cleaners and conditioners. The most commonly used product for leather cleaning is saddle soap. A wipe down with a damp cloth will take care of mildly soiled leather. Be careful not to get the leather any wetter than absolutely necessary.

Vinyl Protectants—After cleaning the seats and door panels, many people like to apply some type of vinyl dressing or protectant. A good compromise is to use a modest amount of dressing. Use a product that is designed for interior use. You want to keep the upholstery supple, but not greasy. You can keep the amount of dressing to a minimum by applying it to a cloth first instead of spraying the protectant directly on the seat. You can also apply the dressing to rubber parts such as the pedals and the weatherstripping.

There are some protectants that contain little or no silicone, and these are recommended. You might also want to check the UV protection qualities of each one as well. Also, some of these products add gloss and protection, but they don't necessarily clean. You may need a separate product for each.

Other Details

Door Panels—Cleaning door panels (especially the front ones) is pretty easy. To thoroughly clean the rear side panels, it helps to remove the seats for better access. Most door panels are vinyl or a combination of vinyl and cloth. Use the same products and techniques for the door panels as for the similar seat materials.

Armrests—Armrests can get quite dirty. They can also trap a lot of dirt around the base edges. There are usually only a couple screws securing the armrest bases to the door panels. For a good cleaning, remove the armrests.

Foot Pedals—Really dirty rubber pedals can benefit from being soaked in a concentrated soap and water solution for a day or so. If the driver's side heel pad is badly worn, you could have an upholstery shop make you a new one, but the cost probably isn't justified. A thorough cleaning and some appropriate vinyl dye (or rubber dressing) can improve the looks of a badly scuffed heel pad.

Wax & Polish Metal Components—Parts of the interior that are painted metal will benefit from a coat of wax. You want a non-abrasive wax, because the paint on these parts isn't very thick. You don't want to rub through to the primer or bare metal.

Chrome or stainless steel trim parts can be kept bright and shiny with a little metal polish. Use the polish sparingly and be careful not to get it on the surrounding upholstery. You can also use the same wax or polish that was used on the painted parts for the chrome trim parts.

Gauge Clusters—The clear plastic gauge and instrument cluster covers should be dusted frequently. These areas seem to be real dust magnets. A small application of plastic polish will keep the lenses in top condition. You can find this polish at paint supply stores, auto parts stores, and at motorcycle shops (it is often used on helmet face shields).

Air Vents—It is difficult to get the dust out of dashboard areas like the air vents. Your best bet is to locate a small vacuum cleaner with very small attachments used for cleaning computer keyboards and other office equipment. You can also loosen the dust with a detailing brush or a cotton

When the ribs are worn off gas and brake pedals, they should be replaced. If there is just a lot of imbedded dirt, soak the pedals in a pan of concentrated dish washing detergent and water. It works best to soak the pedal pads for a day or so.

MINOR REPAIRS & DETAILING

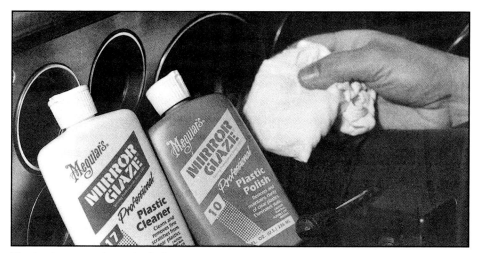

The best way to clean plastic instrument panel lenses is to remove them and wash them. If you don't feel like going to that much effort, Meguiar's Plastic Cleaner and Plastic Polish can make a big improvement in the look of the lenses. These products also work well on colored plastic interior components.

swab. Keep the vacuum running while you loosen the dust. If the vents are easily removed from the dash, it is usually easier to wash them. You will still need to use a small brush to get the corners, but washing is more thorough than vacuuming. Some people apply household dust sprays to the cotton swabs to aid in cleaning vents.

Ashtrays—Another area that needs cleaning on most used cars and trucks is the ashtrays. Remove all debris from the ashtrays and then soak them in a solution of soap and water or a concentrated household cleaner. Use a scrub brush to thoroughly clean the ashtrays. Badly pitted metal ashtrays may need some restoring with sandpaper or steel wool. If you have access to a bead blaster, you can clean up the skunkiest ashtray in a few minutes. The cleaned metal ashtray can be protected with either silver paint or clear lacquer. A plastic ashtray will also look like new with a couple light coats of clear lacquer. It is better to apply a couple clear coats rather than one heavy coat. To help remove any lingering smoke odors, sprinkle some baking soda in the ashtray. Another detailing trick is to whittle some slices off a bar of deodorant soap like Irish Spring and leave the chips in the ashtray.

Remove Stains

Try cleaning the stain with the general purpose cleaners used for seats and carpets. If standard cleaning methods don't produce results, move on to stronger options. There are many commercial products designed for automotive interior stain removal. Carefully read the directions to be sure that the product will safely handle your stain. Many of these products are sprayed on and then they dry to a white powder. The powder is removed and usually the stain goes with it.

Some stains such as non-greasy food can be carefully scraped off cloth upholstery. Use something like a plastic body filler squeegee to gently loosen the dried food from the upholstery fibers. Greasy food can often be removed with a commercial spot remover such as Energine. When working on stains, try to work from the edge to the center. This will lessen the chances of leaving a noticeable ring. After removing a difficult stain, it is a good idea to re-clean the seat or carpet with a general upholstery or carpet cleaner. You don't want the former stain area to stand out from the rest of the material.

Gum—If gum gets in the carpet, try using an ice cube to make the gum harder. You have a better chance of scraping off the gum when it is stiffened. What you can't scrape off, you can try to remove with a liquid spot remover.

Road Salt—Vehicles from parts of the country where the roads are salted in the winter can experience white salt stains on the carpet. These stains can be neutralized with a solution of vinegar and warm water. Dab the solution on the stain and then clean the carpet with regular carpet cleaner.

Tar—Tar or road oil can be a major disaster for carpets. Wax and grease remover, turpentine, or even lighter fluid can be used to remove tar. Just as using tar removers on the exterior of a vehicle requires care, the same caution should be exercised when cleaning carpeting. You want to lightly dab at the stain so you remove the tar with the least possible amount of scrubbing. If you have a choice of tar stains, try to remove the most inconspicuous one first just in case there is any problem with the solvent. Whenever you are in doubt about a nasty stain, take the vehicle to a professional detailer. They encounter these problems on a daily basis so they know what products work and which ones do more damage than good. ∎

INDEX

A

Adhesives, 22
Armrests, 61-62, 132

B

Blades
   electric-edge razor, 16
   hacksaw, 16
   single-edge razor, 16
Bows
   installing, 38-39
   removing headliner, 37-38
Burlap, 82-83

C

Carpets, 43-51, 126
   cigarette burns, 126
   cleaning, 45
   dyes, 118
   installation, 45-51
      cutting holes, 48-49
      final, 48-51
      front section, 49
      padding, 45-47
      securing carpet & padding, 47-48
      trimming excess, 49
   planning, 4-5
   plugs, 126
   quality, 44-45
   removal and floor prep, 45
   trunks, 103
   types, 44
Clamping tools, 18-19
Cleaning
   carpets, 45, 130-131
   door panels, 59
   gauges & trim, 73-75
   interior parts, 28-30
   products, 22-23
   seats, 131-132
   trim, 73-75
   trunks, 100-102
Clip tools, 17-18
Clips & fasteners, 21-22
Clubs (organizations), 23-24
Colors
   application of, 129-121
   changes of, 117-121
      flex agents, 120-121
      matching, 118-119
      touch up, 119
Connectors, electrical, 74-75
Cushions, attaching upper & lower, 89
Custom seats, 112-116
Cutters
   diagonal, 15
   side, 15
Cutting tools, 16-17

D

Dashboards, 67-75, 126-127
   cleaning gauges & trim, 73-75
   electrical connectors, 74-75
   gauge faces, 74
   glove boxes, 75
   minor repairs, 71-72
   modern one-piece, 72
   non-padded, 68-69
   padded, 69-73
      bolt-on pads, 71-72
      partially, 69
   painting, 68-69
   plastic lenses, 74
   plastic parts, 127
      repairing, 127
      touch up, 127
   repairing, 68-69
   repairing cracks, 127
Decals, 101-102
Detailing, 122
   air vents, 132-133
   armrests, 132
   ashtrays, 133
   door panel, 132
   foot pedals, 132
   gauge clusters, 132
   miscellaneous, 132-133
   removing stains, 133
      gum, 133
      road salt, 133
      tar, 133
   tips, 130-133
      cleaning carpets, 130-131
      seats, 131-132
      vacuum frequently, 130
      waxing & polishing, 132
Diagonal cutters, 15
DIY vs. professional tasks, 2-9

Door panels, 52-66
   and armrests, 61-62
   flat-glued, 63-65
      covering, 64-65
      cutting panel boards, 63-64
      drilling holes, 64
      padding, 64-65
   handles & cranks, 54
      moldings, 54
      spacers, 54
   inside the doors, 58
   inspection & reconditioning, 56-57
      insulation, 57-58
      rust-proofing, 58
      water shields, 55-56
      window felt, 57
   installing new, 60-62
   no-sew custom, 62-66
   reconditioning, 58-60
      cleaning, 59
      minor damage of, 59
      molded door panels, 59-60
   removal of, 53-55
      power controls, 55
      rear quarter panels, 55
      retaining clips, 54-55
   replacement, 58-59
   sculptured, 65-66
Doors, 129
   checking alignments, 129
   checking drain holes, 129
   power accessories, 129-130
   repairing weatherstrips, 130
Drills, electric, 16
Dyeing
   vinyl protectants, 120
   which parts are best?, 118
Dyes
   carpet, 118
   and color changes, 117-121
   quality of, 118
   surface preparations, 119-120

E

Electric drills, 16
Electrical connectors, 74-75
Electric-edge razor blades, 16
Equipment, painting, 19-20

INDEX

F
Flex agents, 120-121
Foam
  automotive upholstery, 23
  new, 70
  patches, 124-125
  seat, 83-85
    attaching, 84-85
    lower bolster, 83-84
    type, 84

G
Gas tanks, 101
Gauge faces, 74
Gauges & trim, cleaning, 73-75
Glove boxes, 75
Gum, 133
Guns
  heat, 17
  spray, 19-20, 33
  spray glue, 20
  touch-up, 20

H
Hacksaw blades, 16
Headliners
  installation, 38-40
    bows, 38-39
    miscellaneous items, 42
    notching corners, 40-41
    reinstalling windshields, 41-42
    removing minor wrinkles, 41
    securing edges, 39-40
  removal, 35-38
  removing trim, 35-38
    miscellaneous items, 35-36
    rear package trays, 35-36
    removing headliner bows, 37-38
    removing windshields, 37
    weatherstripping, 37
    windlace, 35
  replacing, 34-42
  securing edges
    front, 40
    rear, 39-40
    sides, 40
Heat guns, 17
Hemmings Motor News, 21-22, 59, 73
Hog rings, 21
  installing, 87-88
  pliers, 14-15
Horns, 128-129

I
Ice picks
Interiors
  painting parts, 26-33
  restorations, 11-12

J
Jute padding, 45

K
Kick panels, 93-97
  covering, 95
  custom, 95-97
  restoring plastic, 94-95
  speakers, 96
  storage pouches, 96-97
Kits
  generic, 3-4
  seat upholstery, 76-92
  upholstery, 3-4
  used, 24-25
  vinyl repair, 125
  where to purchase, 23-25
    clubs, 23-24
    mail order, 24
    retail, 24
    swap meets, 24-25
Knives, 16-17
  putty, 17
  trimmer's, 16-17
  utility, 16

L
Lenses, plastic, 74

M
Mail order, 24
Marking tools, 19
Masking
  dashboards, 32
  interior parts, 31-32
Masking tape, removing, 32
Measuring tools, 19
Mechanics tools, 15-16

P
Package trays, 93, 97-98
  custom, 98
  making a pattern, 98
Padding, 45-47
  cutting, 46-47
  jute, 45
Pads, underlying, 46
Painting
  after restoration, 10
  before restoration, 9-10
  equipment, 19-20
    spray glue guns, 20
    spray guns, 19-20
    touch-up guns, 20
  interior parts, 26-33
    cleaning and rust problems, 28-30
    disassembly, 27-28
    recording everything, 27-28
Paints
  & dyes, 23
  applying, 32-33
    spray cans, 32-33
    spray guns, 33
  basics, 30-33
    masking, 31
    prep, 30-31
    removing masking tape, 32
  spatter, 100
  carpets, 4-5
  color changes, 8
  dashpads, 7-8
  DIY vs. professional tasks, 2-9
  door panels, 6-7
  headliners, 7
  mistakes will happen, 1-2
  partial or complete restoration?, 11-12
  seats, 5-6
  stock vs. custom, 10-11
  trunks, 8-9
  upholstery kits, 3-4
Plastic lenses, 74
Pliers
  hog ring, 14-15
  stretching, 19
Poking tools, 16-17
Power foam saws, 16
Power seats, 116
Products, cleaning, 22-23
Putty knives, 17

R
Reference materials, 20
Repairs
  minor, 122-24
    carpets, 126
Restorations
  interior, 11-12

INDEX

partial or complete?, 11-12
Retail, purchasing at, 24
Rings, hog, 21
Rust problems, 28-30

S

Saws, power foam, 16
Scissors, 14, 16
Scratch awls, 16
Seams, 101
Seat covers
   installing new, 85-92
   reproduction, 77-78
Seat upholstery kits, 76-92
   disassembly, 78-82
     front seats, 80
     removing hog rings, 80-82
     removing seat covers, 80
   finding seats, 79
   installing new covers, 85-92
     attaching upper & lower cushions, 89
     installing hog rings, 87-88
     listing wires, 88
     pivot points, 89
     procedure, 86-92
     removing wrinkles, 89-90
     tricks of the trade, 86-87
   reproduction seat covers, 77-79
   seat repairs, 82-85
     broken springs, 82
     burlap, 82-83
     foam, 83-85
   wrecking yards, 79
Seats, 124-126
   cleaning, 131-132
     cloth, 131-132
     leather, 132
     vinyl, 131
   fading, 125-126
   finding, 79
   foam patches, 124-125
   holes, 125
   power, 116
   using vinyl protectants, 132
   vinyl repair kits, 125
Seats, custom, 112-116
   colors and fabrics, 113-114
     seat swaps, 115-116
       modifications, 115-116
       power seats, 116
       seat base, 115-116
       trucks, 116

   sport, 113
Shears, 14, 16
Side cutters, 15
Side panels, 127-128
Speakers, 96
Spray cans, 32-33
Spray guns, 19-20, 33
Springs, broken, 82
Steering wheels, 105-111, 128-129
   custom, 109-110
     colors & styles, 109-110
     installation, 110
   replacement, 106-107
   restorations
     applying epoxy, 108
     enlarging cracks, 108
     priming & painting, 109
     repairing cracks, 108-109
     sand fillers, 108-109
   steering columns, 110-111
     installation, 111
     repainting and painting, 111
   upgrading, 107-108
Storage pouches, 96-97
Stretching pliers, 19
Supplies, 20-23
   adhesives, 22
   cleaning products, 22-23
   clips & fasteners, 21-22
   foam, 23
   hog rings, 21
   paints & dyes, 23
Swap meets, 24-25

T

Taillights and wiring, 102
Tar, 133
Tools, 13-20
   clamping, 18-19
   clip, 17-18
   cutting, 16-17
   diagonal cutters, 15
   dikes, 15
   heat guns, 17
   hog ring pliers, 14-15
   measuring & marking, 19
   mechanics, 15-16
   poking, 16-17
   and reference materials, 20
   scissors, 14
   shears, 14
   side cutters, 15
   small amount of, 13

   stretching pliers, 19
Torx bolts and screws, 15
Touch-up guns, 20
Trim
   cleaning, 73-75
   removing, 35-38
     miscellaneous items, 35-36
     rear package trays, 35-36
     removing headliner bows, 37-38
     removing windshields, 37
     weatherstripping, 37
     windlace, 35
Trimmer's knives, 16-17
Trucks, seat swaps and, 116
Trunks, 99-104
   carpets, 102-104
   cleaning & repairing, 100-102
   decals, 101-102
   gas tanks, 101
   installing carpets, 103
   minor rust repair, 100-101
   planning one's project, 8-9
   seams, 101
   spatter paint, 100
   upholstery, 102, 104
   wiring & taillights, 102

U

Underlying pads, 46
Upholstery kits
   generic kits, 3-4
   material types, 4
   quality, 4
Utility knives, 16

V

Vinyl
   new, 70-71
   protectants, 120, 132
   repair kits, 125

W

Weatherstripping, 37, 130
Windshields
   reinstalling, 41-42
   removing, 37
Wires, listing, 88
Wiring and taillights, 102
Wrecking yards, 79

Y

Yards, wrecking, 79

About the Author

Bruce Caldwell got his first car, a Garton Kidillac, at the age of five. Forty years and some 200 cars, trucks, and motorcycles later, Bruce is still crazy about cars. He started reading *Hot Rod* in the second grade. Hundreds of model cars were customized in his attempts to duplicate magazine cars. He fully expected a '40 Ford coupe for his eleventh birthday, but was disappointed to receive a more expensive bicycle instead. He has since owned several '40 Fords.

Bruce used his automotive knowledge, an English degree, and a large amount of perseverance to land a staff job at *Car Craft*. He later moved to *Hot Rod*, which was the fulfillment of a childhood dream. He has edited a wide array of enthusiast publications, including *Chevy High Performance, Mustangs and Fords, Custom Painting Annual, Hot Rod Annual, Street Rod Quarterly, Muscle Car Review, Chevy Action* and many others. His how-to articles have appeared in dozens of automotive and truck publications.

Bruce builds his hobby cars and produces articles on them in a well-equipped shop and barn nestled in four acres of large evergreens in Washington state. ■

OTHER HP AUTOMOTIVE BOOKS

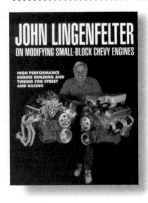   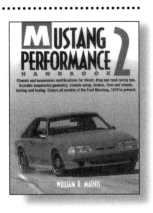

HANDBOOKS

Aerodynamics For Racing & Performance Cars
Auto Electrical Handbook
Automotive Paint Handbook
Auto Upholstery & Interiors
Baja Bugs & Buggies
Brake Handbook
Camaro Restoration Handbook
Classic Car Restorer's Handbook
Engine Builder's Handbook
Metal Fabricator's Handbook
Mustang Restoration Handbook
Paint & Body Handbook
Sheet Metal Handbook
Street Rodder's Handbook
Turbo Hydra-matic 350 Handbook
Welder's Handbook

CARBURETORS

Holley 4150
Holley Carburetors, Manifolds & Fuel Injection
Rochester Carburetors
Weber Carburetors

PERFORMANCE

Big-Block Chevy Performanc
Bracket Racing
Camaro Performance
Chassis Engineering
Chevrolet Power
How to Hot Rod Big-Block Chevys
How to Hot Rod Small-Block Chevys
How to Hot Rod Small-Block Mopar Engines
How to Hot Rod VW Engines
How to Make Your Car Handle
John Lingenfelter On Modifying Small-Block Chevy Engines
Mustang Performance
Mustang Performance 2
1001 High Performance Tech Tips
Race Car Engineering & Mechanics
Small-Block Chevy Performance

ENGINE REBUILDING

Rebuild Air-Cooled VW Engines
Rebuild Big-Block Chevy Engines
Rebuild Big-Block Ford Engines
Rebuild Big-Block Mopar Engines
Rebuild Small-Block Chevy Engines
Rebuild Small-Block Ford Engines
Rebuild Small-Block Mopar Engines
Rebuild Ford V-8 Engines

WEEKEND PROJECTS SERIES

Corvette Weekend Projects
Mustang Weekend Projects (1964-1967)
Mustang Weekend Projects 2 (1968-1970)

GENERAL REFERENCE

Auto Dictionary
Auto Math Handbook
Fabulous Funny Cars
Fiberglass & Composite Materials
Guide to GM Muscle Cars
Turbochargers
Understanding Automotive Emissions Control

TO ORDER CALL: 1-800-223-0510

HPBooks
The Berkley Publishing Group
375 Hudson Street
New York, NY 10014